THE SOCIAL ARCHITECTURE OF GREEN COMMUNITIES

Using case studies from ten European and North American countries, this book discusses the challenges connected to designing viable social architectures in a variety of green communities in urban and rural areas, and in affluent districts and government-subsidised social housing.

Global warming constitutes one of the greatest challenges of our time, but governmental responses have so far been slow and insufficient. Consequently, an increasing number of people have joined a variety of green communities to engage in collective environmental action at the grassroots level. Green communities have the potential to deliver deep and long-lasting behavioural change and carbon footprint reductions, whilst increasing the life satisfaction of participants; however, despite these benefits, they are not spreading as fast as we might expect. Exploring eco-communities, sustainable cohousing, Transition Towns, organic food cooperatives, green neighbourhoods, repair cafés, and more, this book examines why these green communities aren't developing more rapidly, and considers the challenges that they face to their success. By assessing the problems associated with the management of collective life and decision-making, the book provides the reflexive and analytical tools required to better understand the conditions required for their survival and scalability.

Focusing on how the social architecture of green communities contributes to enriching our understanding of social sustainability as a condition for environmental sustainability, this book will be of value to students and scholars across the social sciences, particularly those with an interest in environmental anthropology, sustainability, sociology, public policy, and communication studies.

Quentin Gausset received his PhD in anthropology from the Free University of Brussels and is Associate Professor in Anthropology at the University of Copenhagen. He has 30 years of experience with interdisciplinary research on natural resource management and environmental conflicts, based on fieldwork in Africa, South-East Asia, and Denmark.

Helen Jarvis is a UK based scholar-activist who has enduring interests in the lived realities of communal living, such as cohousing, and grassroots socio-spatial interventions that align social and environmental justice with green sharing economies. Since 2024, Helen has continued to publish and campaign for social and ecological justice as an independent scholar consultant, with Professor Emeritus status.

Simon Lex is Deputy Head of Education and Associate Professor at the Department of Anthropology, the University of Copenhagen. His primary research interests are within climate, organizational and business anthropology. He is a PI of the Grand Solution project "DART: The Danish Model for Citizen Engagement in the Renewable Energy Transition" and co-editor of the *Journal of Business Anthropology*.

Freja Bach Kristensen is a PhD fellow at the Department of Anthropology, University of Copenhagen, and is specialized in the fields of urban anthropology, community development and the green transition. She has extensive experience with civil society collaborations both within and outside academia having worked as a consultant within urban development and citizen involvement for more than 7 years.

Camilla Nielsen-Englyst holds a degree in Educational Studies from Roskilde University and is a grassroots activist rooted in the ecovillage and cohousing movements. For the past decade, her work has focused on connecting research, development, and practice.

Routledge Environmental Anthropology

Environmental Anthropology explores historic and present human-environment interactions, highlighting the link between human-caused environmental problems such as climate change, species extinction, and pollution, with the complex cultural, political, and economic systems that have created them. This series aims to contribute to the growing subfield by providing a comprehensive survey of contemporary topics in environmental anthropology, including food procurement, ethnobiology, spiritual ecology, resilience, nonhuman rights, architectural anthropology, industrialism, and education.

The *Routledge Environmental Anthropology* series welcomes submissions that combine strong academic theory with practical applications, and as such is relevant to a global readership of students, researchers, policy-makers, practitioners, and activists. Please contact Grace Harrison (Grace.Harrison@tandf.co.uk).

Environmental Reflections on the Anthropocene
Nature Transformed
Edited by Gabriel R. Ricci

The Social Architecture of Green Communities
Organizing Grassroots Practivism for the Sustainable Transition
Edited by Quentin Gausset, Helen Jarvis, Simon Lex, Freja Bach Kristensen and Camilla Nielsen-Englyst

For more information about this series, please visit: https://www.routledge.com/Routledge-Environmental-Anthropology/book-series/REA

THE SOCIAL ARCHITECTURE OF GREEN COMMUNITIES

Organizing Grassroots Practivism for the Sustainable Transition

Edited by Quentin Gausset, Helen Jarvis, Simon Lex, Freja Bach Kristensen and Camilla Nielsen-Englyst

Routledge
Taylor & Francis Group
LONDON AND NEW YORK

earthscan
from Routledge

Designed cover image: Getty Images

First published 2027
by Routledge
4 Park Square, Milton Park, Abingdon, Oxon OX14 4RN

and by Routledge
605 Third Avenue, New York, NY 10158

Routledge is an imprint of the Taylor & Francis Group, an informa business

The publication of this book was made possible with support from the Velux Foundation and the University of Vechta.

For Product Safety Concerns and Information please contact our EU representative GPSR@taylorandfrancis.com. Taylor & Francis Verlag GmbH, Kaufingerstraße 24, 80331 München, Germany.

British Library Cataloguing-in-Publication Data
A catalogue record for this book is available from the British Library

ISBN: 9781041293323 (hbk)
ISBN: 9781041293293 (pbk)
ISBN: 9781003769064 (ebk)

DOI: 10.4324/9781003769064

Typeset in Sabon
by Newgen Publishing UK

CONTENTS

CONTRIBUTORS

Sylvia Amar received her PhD in History of Architecture from National School of Architecture Aix-Marseille University. She is an independent curator (Nouvelles Donnes Productions), and an associate researcher affiliated with INAMA Laboratory. Her research explores the spatial practices of intentional communities, ecological architecture, and the cultural dimensions of sustainable living. She has conducted extensive fieldwork in Europe and North America and is currently working on the translation in French and the publication of seminal texts by architect Sim Van der Ryn (1935–2014).

Susan Andreatta, PhD, is a Professor of Applied Cultural Anthropology at The University of North Carolina at Greensboro, USA. Her research and teaching interests coincide with an emphasis on social and environmental justice, food production, health, nutrition, and food security. She leads the sustainable campus garden initiative where she supervises 56 raised beds. Since 2001, she has been the director for Project Green Leaf, a project that was established to assist farmers and the public in sustaining a local agriculture and food system.

Penny Clark is a research fellow at UCL, London, UK, with a focus on climate mitigation and adaptation policy and practice evaluation. Her doctoral research explored and measured the environmental impacts of shared living communities. Penny has edited and contributed to publications about communal living including *Urban Communal Living in Britain,* and *Cohousing in Britain (Volume 2)*. Alongside her academic role, she is a co-founder and director of a consultancy which advises shared living businesses to embed social and environmental impact into their communities.

Ana Margarida Esteves is Coordinator of the Instituto Universitário de Lisboa (ISCTE-IUL), Centro de Estudos Internacionais, Lisboa, Portugal. She has a PhD in Sociology from Brown University and held teaching and research positions in the USA, Belgium, Italy, Brazil and Malaysia. Her scholarship focuses on how place-based responses to the polycrisis, including strategies of commoning, forms of solidarity economy and transition initiatives, organize transnationally to impact economic, cultural and institutional regimes.

Quentin Gausset received his PhD in anthropology from the Free University of Brussels and is Associate Professor in Anthropology at the University of Copenhagen. He has 30 years of experience with interdisciplinary research on natural resource management and environmental conflicts, based on fieldworks in Africa, South-East Asia and Denmark.

Cristina Grasseni is Professor of Cultural Anthropology at Leiden University. She received her PhD in Social Anthropology with Visual Media from the University of Manchester (2001). Her research focuses on food activism, heritage and craft with a multimodal approach. She is co-editor of Elgar's Encyclopedia of Economic Anthropology (2025).

Helen Jarvis is a UK based scholar-activist who has enduring interests in the lived realities of communal living, such as cohousing, and grassroots socio-spatial interventions that align social and environmental justice with green sharing economies. Since 2024, Helen has continued to publish and campaign for social and ecological justice as an independent scholar consultant, with Professor Emeritus status.

Freja Bach Kristensen is a PhD fellow at the Department of Anthropology, University of Copenhagen, and is specialized in the fields of urban anthropology, community development and the green transition. She has extensive experience with civil society collaborations both within and outside academia having worked as a consultant within urban development and citizen involvement for more than 7 years.

Simon Lex is Deputy Head of Education and Associate Professor at the Department of Anthropology, University of Copenhagen. His primary research interests are within climate, organizational and business anthropology. He is a PI of the Grand Solution project "DART: The Danish Model for Citizen Engagement in the Renewable Energy Transition" and co-editor of the *Journal of Business anthropology*.

Camilla Nielsen-Englyst holds a degree in Educational Studies from Roskilde University and is a grassroots activist rooted in the ecovillage and cohousing movements. For the past decade, her work has focused on connecting research, development, and practice.

Anne-Kathrin Schwab is a postdoctoral researcher at the University of Vechta, specializing in transformational and transformative research with a focus on ecovillages, rural transformation, and participatory governance. She holds a doctorate in sociology (magna cum laude) from the University of Siegen and has led and contributed to numerous interdisciplinary research projects at national and European levels. As co-founder *of the International Ecovillage Research Institute*, she bridges academia, politics, and grassroots actors to foster a culture of sustainable change. Her work combines empirical research, policy engagement, and international collaboration to advance socio-ecological transformation.

Ieva Snikersproge received her PhD from the Graduate Institute, Geneva, and is currently an associate researcher at the University of Neuchâtel. Her work focuses on alternatives to capitalism, ecological transition, and post-growth economies. Her latest research examines ecological labour, showing how capitalism transforms human labour in ways that both externalize ecological work and reduce overall opportunities for job creation—thereby creating a systemic addiction to economic growth.

THE SOCIAL ARCHITECTURE OF GREEN COMMUNITIES

An introduction

Quentin Gausset, Helen Jarvis, Simon Lex, Freja Bach Kristensen, Camilla Nielsen-Englyst

The world currently faces an acute climate and environmental crisis grounded in unsustainable levels of consumption and production. While international and governmental action to tackle this crisis has been slow and so far largely insufficient, an increasing number of people organise themselves in green communities to engage in collective environmental action at the local level (Hoff et al. 2016; 2020a; Lockyer & Veteto 2013).

Green communities are groups of people joining forces to live more sustainably, whether it is in eco-villages, intentional cohousing communities, organic food cooperatives, urban gardening groups, green neighbourhoods, transition towns, repair cafés, local energy communities, and so on. All of these communities have an explicit green goal that focuses either on climate (reducing carbon footprint and global warming) or on nature (enhancing biodiversity and reducing environmental pollution), or both (Gausset & Hoff 2024). These communities are intentional: members participate because they want to, not because they need to. In the case of profound disagreement within the community, people can opt out—even though this might sometimes have a high personal cost, such as selling one's housing and relocating elsewhere.

Green communities offer alternative ways of living sustainably that are characterised by more sharing, economies of scale (communities make it possible to benefit from services or infrastructures that are out of reach for individuals), and a strong focus on living in equilibrium with the environment, which can unfold in permaculture design and local organic food production, eco-building, repair and upcycle workshops, self-sufficiency in renewable energy, transport sharing, systems of time banking, among others (Birnbaum 2014; Dawson 2006; Lallement 2024; Lockyer & Veteto 2013; Manzella 2010; Schehr 1997; Sanford 2017). Collective action to reach green goals

DOI: 10.4324/9781003769064-1

translates into environmental behaviour change and significantly lowers carbon footprints, which demonstrates that green communities are an important avenue for the green transition (Clark 2021; Daly 2017; Gausset & Jensen 2024; Lockyer 2017; Sherry 2014, Gausset et al. 2025). Recognising the determining role of collective action on behaviour change constitutes a rupture from conventional psychological and economic theories that focus on influencing individuals through nudges or economic incentives (Gausset & Hoff 2024).

But communities also come with great challenges. Members must agree on collective action and successfully manage conflicts (Palgi & Getz 2025; Pickerill 2025; Shenker 2011). Communities take time to build—people must learn to know and trust one another and design efficient and inclusive ways of making decisions. In their search for a *modus vivendi* and a *modus operandi*, green communities develop alternative types of social and political organisations—what we call here alternative forms of 'social architecture'. Yet, despite the importance of social architecture for green communities and the green transition, this aspect of green communities remains understudied.

The different chapters of this volume contribute to filling this knowledge gap by presenting a collection of green communities located in Europe and North America that focus on sustainable housing, regenerative agriculture, short circuit organic food markets, repairing and upcycling, or renewable energy, among others. We discuss how green communities are created and reproduced, how they plan for community transformation, how they provide the motivation for behavioural change, how they interact with municipalities and other state institutions, how they manage time and common resources, how they address social differences, and how they make decisions and solve conflicts. By describing the social architecture of green communities, the 11 chapters of this volume contribute to a better understanding of how alternative social organisations can contribute to the green transition.

Environmental and social impact

Green communities have an important environmental and social impact. Different surveys have found that the carbon footprint of members of green communities is much lower than in the rest of the population. A comparison of 141 green community members in Italy, Germany, Romania, and Spain with 1476 non-members sampled from the same geographical regions showed that the carbon footprint of members of green communities is 16% below average and demonstrates at the same time a significantly higher life satisfaction (Vita et al. 2020). Another study comparing 255 members of green communities in Denmark with 1018 representative Danes showed that members of green communities have a consumption-based carbon footprint that is 28% below average at the same time as they have a higher life satisfaction (Toft et al.

2021). In that same survey, the lowest carbon footprint, 60% below average, was found in an eco-village in the Funen region and was still accompanied by a significantly higher than average life satisfaction (Gausset & Jensen 2024).

A literature review of 16 intentional communities in the world shows that the average carbon footprint in intentional communities is ~35% less than the comparison figures (Daly 2017: 1368). Most recently, a survey comparing the consumption-based carbon footprint of 1002 representative Danes and 505 members of Danish green communities showed that the average carbon footprint of members of green communities is approximately 30% below average for people living in cohousing and members of the society for nature and conservation, and around 45% for members of green neighborhoods or eco-villages (Gausset et al. 2025). Once again, the life satisfaction of members of green communities is higher than average. A range of other studies confirm these results in a variety of contexts (Clark 2021; Grinde et al. 2018; Lietaert 2010; Lockyer 2017; Mulder 2005; Sherry 2014). The over-riding conclusion of this summary of surveys is clear: collective action in green communities leads to dramatic reductions in carbon footprints.

The higher life satisfaction and the lower carbon footprints found in green communities are a remarkable result that contradicts various studies showing that happiness correlates positively with wealth and therefore with consumption and a higher carbon footprint. These studies also show that people living in wealthy nations are statistically happier than in poorer nations and within nations, wealthier citizens are happier than poorer citizens (Duncan 1975; Easterlin 1973, 1995, 2001; Lane 1993; Lee 2005; Wang et al. 2017). The fact that members of green communities display higher levels of life satisfaction at the same time as they are characterised by lower average income, lower consumption, and lower carbon footprints suggest that factors other than wealth and materialist consumption are also at play in determining happiness. For example, these factors can be a higher social equity (Brockmann et al. 2009; Csikszentmihalyi 1999; Oishi et al. 2011; Wilkinson & Pickett 2009), more leisure, especially when coupled with contentment (Csikszentmihalyi 1999; Sahlins 1972; Jackson 2005), a richer social life and less loneliness (Ortiz-Ospina & Roser 2020; Satici 2016) or being able to live up to societal and personal norms to avoid anomie and cognitive dissonance (Festinger 1957; Durkheim 1897).

Social architecture

Since the Brundtland report (1987), sustainability is commonly described as having three pillars. Environmental sustainability is a question of keeping natural resources and ecosystems in balance, economic sustainability refers to a stable economy that supports both the environment and society, and social sustainability is described in terms of social justice, which ensures that everyone thrives. Securing these three pillars requires a complex social organisation to determine how resources are owned and managed, how citizens negotiate decision making, and how they share the burdens and the benefits of resource management. In other words, achieving sustainability requires a social architecture that works for this purpose. Studying different cases at the local scale allows researchers to grasp the complex social interactions, relations, and arrangements that are conducive to achieving sustainability. Each local green community described in this book represents a certain path, a certain social architecture, and a certain way of achieving a dynamic equilibrium between these components.

The environmental impact of green communities is achieved through a variety of collective actions. These include living in smaller homes and sharing bigger collective spaces, sharing or borrowing material objects instead of buying new ones, and having economies of scales to allow members of the group to have access to services that are out of reach for individuals. These services could be a vegetable box from local food producers, local district heating, an extensive waste sorting infrastructure, shared cars, collective gardening fields, rearing farm animals, collective meals, sustainable energy production, ecological sewage system, collective infrastructure, etc.). All these forms of services not only require a physical infrastructure comprised of buildings, land, machines, tools, roads, transport means, wind turbines, and so on. They also require a social infrastructure that organises collective action for the green transition and equally important, to affect individual behaviour change (Nguyen 2019; Gausset & Jensen 2024).

Since the term 'social infrastructure' can be mistaken as referring to buildings and facilities designed to promote social interaction (see, for example, Brown & Barber 2012; Davern et al. 2017; Grum & Grum 2020), we prefer to use the concept of social architecture to talk about the organisation of labour and mutualities that enable the everyday functioning of green communities and their physical infrastructures (Jarvis 2015). The social architecture of green communities comprises working groups, systems of time banking, and systems of sharing and borrowing material items. It also includes written and unwritten rules, values, norms, and the formal and informal institutions that contribute to the effective organisation of the community.

Talking about social 'architecture' presupposes that it is purposefully designed, even if the design is the result of muddy collective action rather than the brainstorm of a genius. Talking about architecture also entails a certain solidity and permanence while it can also be constantly transformed, augmented, and enriched. Architectures are human-made, collaborative, and the result of a long chain of decisions and actions.

The green communities discussed here are diverse and range from eco-villages or sustainable cohousing communities (in chapters authored by Amar, Clark, Esteves, Gausset & Englyst-Nielsen, Jarvis, Schwab, and

Snikersproge) to urban gardening communities (Andreatta and Grasseni), energy communities (Lex), and green communities in local neighbourhoods or in social housing (Gausset & Englyst-Nielsen, Kristensen). These green communities tend to be small and based on grassroots engagement at the local level. They work with a flat and informal political structure, unlike environmental associations that are more formally structured and include a larger number of members. Some green communities hold important resources in common, as in cohousing, which lead to investing time and effort to foster inclusive decision making. Other green communities share no common resources and require little commitment from members, which makes these communities more volatile and vulnerable to social fluctuations. The main point here is that no two green communities are alike, which is evident in the descriptions of the book's different case studies, and every single community discussed here has developed its own social architecture through its own history and through the influence of its members.

The resilience and liveability of green communities depend on the way their social architecture is designed. A community is certainly more than the sum of its individuals, but being a member of a community also has a cost in terms of the time and resources invested in designing and maintaining its social architecture. Green communities do not emerge from an *a priori* categorial identity; rather, they are created by sustained and coordinated efforts towards more sustainable practices (a joint commitment) while fostering feelings of togetherness (or affective belonging) (Amit 2010). However, this process is not tension-free. Green communities are riddled with conflicts big and small; finding consensual forms of living together is key to their resilience. Understanding the social, political, and economic arrangements that hold the community together can yield important lessons for the green transition.

Practivism

The green communities discussed here are 'practivist' in the sense that the members of these communities focus on changing their own practices instead of trying to change those of others (as in activism). Practivism is a form of activism in that it opposes the mainstream status quo and supposes action that triggers change. But whereas classical activism is oriented outward and aims to change others, practivism is oriented inward and aims to change oneself (Hoff et al. 2020b).

Activism takes its roots from protest movements of citizens acting against public policies that are detrimental to the environment and trying to influence politics and society through demonstrations, petitions, direct actions, and lobbying (Lubell 2002; Rootes 2003; Wapner 1996). Environmental activism, more specifically, has a long history and has known different iterations, from the creation of organisations that aim to preserve biodiversity to the establishment of the anti-globalist movement (Connolly et al. 2012; Dalton 1994; Gausset & Hoff 2024). Practivism, by contrast, can be said to take its roots from cognitive dissonance (Festinger 1957), i.e., in action to remove the mental discomfort deriving from a mismatch between a person's values and behaviour. The drive to build green practivist communities is grounded in the frustration of living unsustainably in a mainstream society that makes little room for alternative and more sustainable ways of living. People acting on this frustration engage in building green practivist communities that become parallel micro-societies that facilitate sustainable living through collective action organised within a specific social architecture.

Although activism and practivism have different roots and goals, they remain complementary forms of engagement for the green transition. One finds practivists who shy away from activism and are reluctant to even sign a green petition; and one also finds green activists that have a high carbon footprint and make little effort to reduce it. But there is an overlap: one also finds people who are both activists and practivists and consider both postures as two sides of the same coin.

While activists tell other people what they should do, practivists focus on doing it and very consciously and explicitly avoid 'moralising' or 'pointing their finger' at other people (Gausset 2020; Gausset & Hoff 2024). If practivists influence others at all, it is by being a role model or inspiring others, not by giving normative moralising discourses. Green communities wield influence on their mainstream neighbours because their initiatives can serve as a source of inspiration for others and encourage them to embrace similar environmentally friendly practices (Gausset & Jensen 2024; Gausset et al. 2025). Green communities actively work to establish models of sustainability that resonate beyond their boundaries and prompt discussions about what it means to be part of the larger ecological and social community.

The green communities discussed here are practivist in that their members organise at the grassroots to adopt more sustainable practices and improve their local environment. These cases range from tightly knit communities sharing many things in common and having daily interactions between members (for example, in eco-villages and other intentional communities) to loosely knit communities that have no common resources and less intense or less frequent social interactions (as, for example, in green neighborhoods, transition towns, food cooperatives, and the like). They all provide sources of inspiration for other communities that would like to live more sustainably.

Idealism and pragmatism

Most people collaborating to design and organise green communities are motivated by the idea of a better, more sustainable world. But they do not all share a utopian mindset. There are people and groups that exert a practical understanding of how to implement sustainable practice that is based on a theory of meaningful change grounded in the place where they live, and there are those who consciously or unconsciously embrace a utopian perspective. Of course, utopian ideals and practices are contested and nuanced and exist on a continuum. For example, many newly forming groups and communities visit and learn from practical utopias (iconic intentional communities): they 'see for themselves' that another world is possible. For this anthology, we find the concepts of utopian practice, everyday utopias, and a recent turn toward prefiguration (Wright 2020) to be helpful in strengthening our case for focusing on green community practivism.

Lyman Tower Sargent (2010) concisely outlines 'three faces of utopianism': literary utopia, utopian practice, and utopian social theory. Utopian practice includes a wide range of intentional communities as well as new expressions of sharing and solidarity. While the vision and values of 'social dreaming' mean different things to different people, Ruth Levitas (2013) argues that 'the desire for a better way of being' should be understood as a process or method of thinking differently, as an 'education of desire', rather than a prescribed ideology or end-goal. When placed beside distinctions between utopia and utopian methods, intentional communities can be seen as complex phenomena with long histories occurring in many different settings (Levitas 2008). Because intentional communities differ radically from time-to-time and place-to- place, we need to delve deeper into their social architecture to capture paradoxical combinations of imagined and realised continuity and change.

Lucy Sargisson (2012) and others claim that utopian practice maintains a philosophy of hope that is essential to any vision of a better future. This quality of hope provides the impetus for activists to challenge and change the status quo and pioneer alternatives to mainstream society that counter the unhelpful feelings of powerlessness that shape the way many people respond to climate anxiety. For example, Ernst Bloch et al. (1986: 7) conceptualise utopia as 'the forward dream' and what is 'not yet' with the emphasis on 'yet' expressing hope in possibility. This is what many green groups need to believe in to keep going and establish their green initiatives. Yet, critics observe that 'social dreaming' does not take us very far (or far enough) as a theory or a means of achieving meaningful change. Utopian practice can remain stuck at a stage of anti-establishment opposition and dissatisfaction because it lacks effective tools and tactics and a practical guide to change.

In response to this critique, recent trends in degrowth activism appear to replace utopian ideals with a prefigurative politics. Prefiguration is a concept originating from historical social movements, particularly anarchism, that involves living 'as if' a future scenario is the reality in the here and now (Cooper 2020). Prefigurative techniques of future visioning are widely used to test alternative and potentially scalable models (degrowth, solidarity, shareable commons, etc.) and develop internal processes and built structures that accommodate transitions beyond market economies and polities (Nelson & Chatterton 2022: 60). Davina Cooper defines these 'networks and spaces that perform regular daily life in a radically different fashion' as everyday utopias (Cooper 2013: 2). During the COVID pandemic in the early 2020s, for example, many hyper-local mutual aid groups prefigured forms of solidarity between neighbours by reaching out to isolated and vulnerable citizens. Prefigurative forms of solidarity are also widely credited for the way that nascent intentional communities imagine and practice solidarity before they are physically able to occupy a common house, i.e., by cooking and sharing meals together, carpooling, and learning how to make collective decisions.

Critics observe that prefiguration tends to function as a metanarrative that reflects a privileged point of view because it is largely shaped by those whose concern for the future are not rooted in fighting to survive in the present (Pellizzoni 2020; Törnberg 2021). Notwithstanding that the social dimensions of solidarity and uneven commitment to practising alternatives are underexplored, prefiguration shows that activists 'will weave together practices and values that will sometimes feel embedded or trapped in capitalist ways of doing things and at other times will be more liberatory or antagonistic' (Chatterton & Pickerill 2010: 488). Prefigurative activists practice the future they wish to see in open and experimental ways that 'have the potential to generate solidarities' (Chatterton et al. 2013: 611). In this way, it is on the pragmatic end of an idealist-realist continuum.

By presenting an alternative way of doing things, these everyday utopias enable a de-familiarisation of taken-for-granted assumptions and practices that enable a critical engagement with society (Cooper 2020). Although similar in many ways, prefigurative future visioning (living 'as if' a future scenario is the reality) aligns more closely with a theory and strategy of structural and political change, while practivism is less frequently seen as challenging existing power relations or as an existing threat to the system as a whole.

The chapters in this book describe different forms of prefigurative politics, whether in trying to develop sustainable housing communities, alternative systems of food production and marketing, green neighbourhoods, or energy communities. The green communities described in this anthology work with high ideals of solidarity, inclusion, justice, and sustainability, sometimes more in theory than in practice (as discussed in chapter 2) but most often when trying to find new forms of pragmatically balancing the different aspects and designing new forms of social architectures into the process.

Scalability

There are tensions and contradictions in the 'scalability' of eco-social housing and community transformations. The scale and focus entailed in people building their own homes can appear self-interested and parochial against a backdrop of global crisis. Yet as Nabeel Hamdi (2004) cogently observes in his book *Small Change*, big ideas often come from small places and the human scale of social change can have a huge positive multiplier effect. This partly explains why intentional green communities are often regarded as a 'living laboratory of how the world could be different' (Chatterton 2015: 17).

Small-scale projects pose a challenge for diversity (income, gender, generation, ethnicity and lifestyle). Much as we define the notion of 'economies of scale' through a whole and circular economic lens and acknowledge multiple, diverse, and non-market solidarity economies, we understand that what it means to grow green communities 'at scale' will involve a combination of different scaling processes.

From a review of the literature, we identify four scaling processes: scaling up (promoting an institution's shifts in law, policy, and resource flow to replicate a proven concept), scaling out (proliferating the core principles and approaches to other places and projects), scaling deep (change values and mindsets and engage with a theory of change to increase impact beyond the initiative via durable system-wide change), and scaling within (strengthen and embed initiatives in their context, making them more resilient to upcoming challenges) (Moore et al. 2015; Lam et al. 2020; Omann et al. 2020). In local and national governments, scalability is usually interpreted through 'scaling up' and 'scaling out' prototypes that are easily replicated. By contrast, non-market players thrive by remaining small and low-cost. In this context, scalability is better understood by what it takes to retain excited and engaged people, scaling deep and within. We argue that scaling up or out is insufficient to translate into durable systemic changes if they are not supported by cultural amplification and social learning processes.

There are compelling arguments for encouraging multiple routes and scales of testing the viability and scalability of niche green groups and projects because the contemporary global landscape is increasingly defined by a convergence of interlinked and mutually amplifying crises: climate catastrophe, economic precarity, polarised communities, geopolitical conflict, displacement, and the erosion of democratic institutions. When scaling out, care should be taken to align sustainability with fairness and with new forms of democratic participation.

For example, as outlined in chapters 6 and 8, intentional green communities scale out by receiving financial and organisational support from regional municipalities, non-profit organisations, academic institutions, and private consultants. These cross-sector interactions expand social infrastructures through coordinated efforts, shared resources, and collaborative planning, which enable a wider adoption of sustainable practices and policies. Moreover, these interactions support the exchange of knowledge and success stories, which serves as inspiration for new communities and grassroots initiatives to emerge. As a result, the environmental, social, and economic benefits of intentional green communities go beyond individual neighbourhoods: they exert influence on neighbouring regions and national political agendas.

Engaged auto-ethnography

A striking feature of the different chapters of this book is that most authors are engaged daily in a local green community not just as researchers but also as practivists. Whether anthropologists, sociologists, geographers, or architects, these authors derive their knowledge from long-term living and/or participating in an intentional housing community, a local urban gardening project, or a local green neighbourhood. Even though their experience sometimes comes from formally planned research projects, it is often supplemented by sheer substantial personal life experience as citizens engaged in the green transition.

Therefore, the research methods used in many chapters go beyond classic participant observation in which an external researcher 'goes native' and participates in a foreign community that is unknown to them in order to observe how it is organised. Instead, they rely more on 'observational participation' (Gatt & Ingold 2013) and on auto-ethnography, in which participation comes first and naturally as part of one's personal experience in struggling to 'make things work', finding solutions, and moving forward. Observation and rationalisation come at a later stage, when comparing one's experience with that of others, trying to help other communities in solving their own challenges, and when using inductive theorising that is grounded in personal social relations.

In this book, engaged auto-ethnography derives from personal and often long-term engagement in the green transition that occurs prior to and is independent of conducting research and collecting data. It therefore goes beyond action research, which is classically understood as the temporary involvement of a researcher in contributing to the solving of the problem under study for the sake of collecting higher quality research data. In engaged auto-ethnography, the researchers use their academic experience studying others in past research projects in order to understand their own situation in a current local project borne by a local green community and theorise their own personal experience as members of these communities. This long-term personal engagement provides a wealth of remarkable case studies, fine and subtle observations, and a holistic perspective. Most importantly, researchers find themselves at the core of social interactions in a privileged position to understand the struggles and mechanisms that generate societal transformation. Fully aware that they are living in the Anthropocene, these authors engage as citizens in the green transition and use their research skills to learn from their own experiences in order to draw generalisable lessons from them.

Community building and diversity

While the design of social architecture takes form and happens at the grassroots level, it can also be strongly influenced by governmental authorities, whose regulations shape and put limits on the realm of the possible. The higher the level of resources involved in community building, the more external players (municipal entities, housing estates, energy developers, building contractors) influence its design. As green communities relate to external stakeholders, they raise critical questions regarding boundaries. How do established community borders influence cooperation and resource sharing? What role do these boundaries play in defining community identity and fostering or hindering collective action? When we address such questions, it becomes clear that the interactions between intentional communities and their surroundings are not only about environmental practices, they are also about redefining the very nature of community itself. Through these collaborations, intentional communities can explore and negotiate their boundaries, which can lead to an approach to green community building that reflects both local and regional needs.

Influence 'from the top' can have good intentions and can be seen as supportive of citizen action, which adds an element of positive alignment. But it is also a way to steer the engagement of citizens in certain directions that are politically prioritised. As discussed in chapters 5, 9, and 10, municipalities and housing associations do not simply have a green agenda. They also have social and political agendas for their residents: social inclusion, compliance to renewable energy plans, or the instrumentalisation of volunteer work to supplement welfare provisions. They also have personnel trained to mobilise, motivate, and initiate or steer green and socio-political initiatives, which are often seen in combination. As the cases discussed in these chapters demonstrate, this can trigger resistance at the grassroots (sometimes expressed as apparent apathy while using the power of 'non-participation' to resist), but it can also lead to creative strategies that maximise the benefits of collaborating with external actors while minimising their influence on the social architecture. Such interactions emphasise the dynamic nature of community building, which are influenced by ongoing negotiation and redefinition, both internally and with external actors, and ultimately playing a vital role in shaping both community identity and the collaborative efforts towards sustainability.

While green communities share some fundamental characteristics everywhere, they also evolve in different national and cultural contexts that shape differences among them. The form of the State, including its level of centralisation or decentralisation, its trust or mistrust of civil society, and the level of latitude and freedom it gives to citizens and communities, influences its positioning vis-à-vis green practivist communities and its interaction with them. For their part, green communities sometimes emerge in opposition and resistance to the state, sometimes in close collaboration or even as an extension of the state. The historical role played in society by cooperatives, civil society, countercultural movements, and the balance between collectivism and individualism take different forms in eastern, northern, and southern Europe or in North America. By the same token, social class, equity, cohesion, inclusion, exclusion, and leadership take on different colours and flavours in different countries.

Case studies

The case studies in this book represent a sampling of these large and diverse forms of green communities. The following is a summary of how they contribute to enriching our knowledge of the social architecture of green practivist communities.

This first chapter by Helen Jarvis applies a multi-level approach and a theory of change to situations of dispersed placemaking in the UK. Comparisons are usefully drawn between the 'ordinary' activism of people coming together to repair and improve ordinary places and the 'communal' activism of intentional eco-communities where resident members practice a shared vision of how to live well together. Five hyper-local placemaking case studies are presented from the UK cities of London, Birmingham, and Newcastle upon Tyne to illustrate how green groups and communities can retrofit a sense of place and purpose within existing social and built environmental systems. The author's discussion and analysis highlight the attributes of interpersonal connections (people-power) such as trust and purpose as the 'glue' that drives collective action and effective change. Jarvis also identifies these as crucial factors for the viability and scalability of niche sustainability transformations. The case studies provide illustrative rather than representative insights into the ways that niche innovation occurs in hyper-local networks of practivist actors who consciously interact in shared spaces of togetherness. This process is not always recognised by policy makers, but it is possible for early adopters of social innovation to pass it on to a larger population and influence the values and norms of institutions.

The second chapter, by Ana Margarida Esteves, compares two green intentional communities, Tamera in Portugal and Monte Verità in Switzerland, that share overlapping cultural aims and represent a similar pursuit of cultural innovation and social regulation in different locations and historical contexts. The chapter's ethnographic analysis explores how intentional communities that prioritise cultural innovation over structural change deploy counter-cultural forms of social regulation as a form of soft power. Esteves argues that social inclusiveness is a prerequisite for the scalability of green communities. A community dominated by core members who come from the more affluent sectors of mainstream society often discourage the participation of individuals who do not share their cultural and economic background. Dissolving cultural and economic barriers faced by diverse populations who want to access green intentional communities is key to the growth of green communities.

In the third chapter, Quentin Gausset and Camilla Englyst-Nielsen compare three Danish green communities and describe how they manage conflicts and make collective decisions. Green communities are associated with drastic reductions in carbon footprints, but they are not free from conflicts and face many difficulties when they try to make decisions for the community. Both majority rule and consensus, two traditional decision making methods, can split communities into 'winners and losers', and neither is immune to being highjacked by individuals. The authors discuss three Danish case studies that describe the micro-politics in\ an eco-village, a green neighbourhood, and a bee-keeper guild. The chapter demonstrates how green communities can avoid the problems linked to consensus and majority rule by using innovative forms of micro-politics such as sociocracy and do-ocracy, which allow them to make decisions that are more efficient, more agile, more inclusive, and less prone to conflict.

The fourth chapter by Ieva Snikersproge draws on fieldwork in the French 'back-to-the land' movement. While it is widely assumed that any alternative to capitalism must be 'collective', what collective means remains open to debate. In order to understand the conditions of participation, collective practices, and decision making and to explore the scalable potentials for ecological transformation, this chapter presents a typology of green communities in France, including a commune, an eco-village, an eco-habitat, and alternative networks of individual households. Scalability requires accessibility, but collectivisation is hindered by the cost of accessing property, which cannot be offset by subsistence activities or economy and which can ultimately create conflicts about how the different ways of contributing to the community are valued.

Simon Lex describes in chapter 5 how large-scale, top-down renewable energy projects implemented in three Danish municipalities in the Jutland Peninsula threaten to destroy local communities and local environmental balance, which can lead to fierce, local resistance. However, contrary to the perception of opposition driven by NIMBY ('not in my backyard') individualists, this resistance is often organised collectively and is not opposed to renewable energy projects per se. Instead, it aims to protect local communities and environments. By introducing the concept of constructive disobedience, this chapter presents an analytical tool designed to understand how local communities mobilise and develop alternative solutions that not only align with national climate targets but also reflect local cultural and environmental values and contexts.

In chapter 6, Anne-Kathrin Schwab demonstrates how participatory methods used by eco-village and transition town movements can stimulate the scalability of sustainability transitions in existing rural and 'rurban' communities. By combining trust-building, inclusive participation, culturally sensitive facilitation, and empowerment strategies, it is possible to catalyse community reflection, i.e., the co-creation of visions and practical steps toward both environmental and social sustainability. This contributes to the broader debate on transformative science by showing that participatory methods originally developed in intentional communities can be used to design and foster change in existing communities, provided these methods are introduced with respect for existing traditional neighbourhood structures and informal governance patterns.

Penny Clark explores in chapter 7 the practical and social mechanisms of sharing in three UK intentional communities. Given that a significant portion of domestic greenhouse gas emissions is generated by energy consumption at home, sharing in and across households can provide significant environmental benefits. Through the lens of social practice theory and social norms, this chapter examines how spaces and objects are shared in three intentional cohousing communities in the UK. The chapter demonstrates that using workplace practices such as meetings, explicit rules, and management systems help manage the shared domestic space. The chapter further argues that making these sharing practices work requires a reconceptualisation of what 'privacy' and 'home' mean and calls for continuous emotional and relational labour. These competences and skills are necessary for increased sharing and offer important lessons for other intentional communities.

Discussing four case studies from France, Italy and Denmark, Sylvia Amar advances in chapter 8 the concept of 'empathic design' to account for how contemporary eco-villages develop their spatial and social organisation as concrete responses to today's ecological and social challenges. It argues that by favouring architectural practices that are adapted to local spaces and subsistence economies that are grounded in values of simplicity and resilience, eco-villages embody alternative forms of inhabiting domestic space based on the principles of shared governance and empathy. Such eco-villages develop ways of life that combine local autonomy with land use and regional integration. Empathic design is a key aspect of working in eco-villages as laboratory spaces that are capable of inspiring the fields of contemporary architecture and land use planning that are oriented toward ecological and solidarity-based planning.

In chapter 9, Freja Bach Kristensen describes how green community building is increasingly promoted as a viable pathway to foster more environmentally friendly lifestyles. Her chapter explores how this goal is implemented in the context of social housing in Denmark and argues that promoting social and green activities is not always sufficient in and of itself to build liveable green communities in social housing estates. The success of green communities depends on the engagement of key local individual brokers, who take on the practical and mental labour of green community building. By applying the concept of sustainability brokering (the translation of wider notions of sustainability into locally attuned activities that can mobilise other residents), the chapter highlights the crucial role of these individuals and the importance of their efforts. It also outlines the skills, competencies, and the general organisational and structural conditions necessary to promote these efforts.

Based on fieldwork in a community garden in the Netherlands, Cristina Grasseni describes in chapter 10 how sustainable food production is local, hand-made, crafted, and slow, and therefore often more expensive than large-scale production, which makes it an easy target of municipal budget cuts in times of crisis. Frictions emerge from the confrontation between volunteers and governmental actors. Practivist and craftivist green volunteers are interested in learning new skills at the grassroots level and establishing alternative food citizenships. On the other hand, municipal or other institutional agents try to impose top-down bureaucratic, productivist, and contradicting sustainability narratives that stifle the motivation that generally animates volunteers. The chapter also describes the dependency that many community gardens have vis-à-vis local municipalities and institutions to access land and financial support and how collaboration between both these parties is impeded by opposing values. Because green communities need institutional support and institutions rely on green communities to engage citizens in the green transition, both these partners need to develop common ground for on-going and sound collaboration.

Lastly, Susan Andreatta highlights in chapter 11 the lessons to be learned from a community garden that has coexisted with a US university since 2010. It discusses the social architecture of engagement and encounter created by students, faculty, staff, alumni, and local residents, who plant organic food together as a green community dedicated to the education and well-being of its volunteer members. This case study discusses the role that community gardens and food sharing can play in local food systems. It provides ethnographic evidence of factors of social experiential learning and a practivist approach that is crucial for environmental awareness, motivation, and enduring grassroots governance over the long run. While the chapter's empirical insights describe how the University of North Carolina Greensboro Campus Garden and its student volunteers learn how to grow produce, herbs, and flowers, the analysis and lessons learned from this enterprise have wider relevance. In order to cultivate and maintain grassroots contributions to liveable and sustainable local communities, the basic questions of viability and scalability call for the diffusion and normalisation of the shared learning architecture. This social architecture embeds the critical component of collective engagement in both formal environmental education and informal social experiential learning.

References

Amit, V. (2010). Community as 'good to think with': The productiveness of strategic ambiguities. *Anthropologica*, 52(2), 357–363. https://www.jstor.org/stable/29546038

Birnbaum, J. (2014). *Sustainable revolution: Permaculture in ecovillages, urban farms, and communities worldwide*. North Atlantic Books.

Bloch, E., Plaice, N., Plaice, S., & Knight, P. (1986). *The principle of hope*. Vol. 3, 1954–1959. MIT Press.

Brockmann, H., Delhey, J., Welzel, C., & Yuan, H. (2009). The China puzzle: Falling happiness in a rising economy. *Journal of Happiness Studies*, 10, 387–405. doi: 10.1007/s10902-008-9095-4.

Brown, J., & Barber, A. (2012). Social infrastructure and sustainable urban communities. *Engineering Sustainability*, 165(1), 99–110.

Brundtland, G.H. (1987). *Our common future: Report of the World Commission on Environment and Development*. United Nations.

Chatterton, P., Featherstone, D., & Routledge, P. (2013). Articulating climate justice in Copenhagen: Antagonism, the commons, and solidarity. *Antipode*, 45(3), 602–620.

Chatterton, P. (2015). *Low impact living: A field guide to ecological, affordable community building*. Routledge.

Chatterton, P., & Pickerill. (2010). Everyday activism and transitions towards post-capitalist worlds. *Transactions of the Institute of British Geographers*, 35, 475–490.

Clark, P. (2021). *Practices of shared living: Exploring environmental sustainability in UK cohousing, community living, and coliving*. [Doctoral dissertation, University of Westminster].

Connolly, J., Smith, G., Benson, D., & Saunders, C. (2012). *Politics and the environment. From theory to practice*. Routledge.

Cooper, D. (2013). *Everyday utopias: The conceptual life of promising spaces*. Duke University Press. https://doi.org/10.1515/9780822377153

Cooper, D. (2020). Towards an adventurous institutional politics: The prefigurative 'as if' and the reposing of what's real. *The Sociological Review*, 68(5), 893–916.

Csikszentmihalyi, M. (1999). If we are so rich, why aren't we happy? *American Psychologist*, 54(10), 821–827. https://doi.org/10.1037/0003-066X.54.10.821

Dalton, R. (1994). *The green rainbow: Environmental groups in Western Europe*. Yale University Press.

Daly, M. (2017). Quantifying the environmental impact of ecovillages and co-housing communities: A systematic literature review. *Local Environment*, 22(11), 1358–1377.

Davern, M., Gunn, L., Whitzman, C., Higgs, C., Giles-Corti, B., Simons, K., … Badland, H. (2017). Using spatial measures to test a conceptual model of social infrastructure that supports health and wellbeing. *Cities & Health*, 1(2), 194–209. https://doi.org/10.1080/23748834.2018.1443620

Dawson, J. (2006). *Ecovillages: New frontiers for sustainability*. UIT Cambridge.

Duncan, O.D. (1975). Does money buy satisfaction? *Social Indicators Research*, 2(3), 267–274. https://doi.org/10.100.7/BF00293248

Durkheim, E. (1897). *Le suicide. Etude de sociologie* [Suicide. A sociological study]. Alcan.

Easterlin, R.A. (1973). Does money buy happiness? *Public Interest*, 30, 3–10.
Easterlin, R.A. (1995). Will raising the incomes of all increase the happiness of all? *Journal of Economic Behavior & Organization*, 27(1), 35–47. https://doi.org/10.1016/0167-2681(95)00003-B
Easterlin, R.A. (2001). Income and happiness: Towards a unified theory. *Economic Journal*, 111, 465–484. https://doi.org/10.1111/1468-0297.00646
Festinger, L. (1957). *A theory of cognitive dissonance*. Stanford University Press.
Gatt, C., & Ingold, T. (2013). From description to correspondence: Anthropology in real time. In W. Gunn, T. Otto, & R.C. Smith (Eds.), *Design anthropology: Theory and practice* (pp. 139–158). Bloomsbury.
Gausset, Q. (2020). Stronger together. How Danish environmental communities influence behavioural and societal changes. In J. Hoff, Q. Gausset, & S. Lex (Eds.), *The role of non-state actors in the green transition: Building a sustainable future* (pp. 52–70). Routledge.
Gausset, Q., & Jensen, P.D. (2024). Living sustainably in a Danish eco-community: How social and physical infrastructures affect carbon footprints. *npj Climate Action*, 3, 33. https://doi.org/10.1038/s44168-024-00113-5.
Gausset, Q., & Hoff, J. (2024). Fællesskab – Et overset element i den grønne omstilling [Community – An overlooked element in the green transition]. In S. Krøijer, & L. Tønder (Eds.), *Grøn samfundsteori: Tanker til en økologisk krisetid* [*Green social theory: Thoughts for an ecological time of crisis*] (pp. 367–83). Gyldendal.
Gausset, Q., Suszkiewiecz, N.J., & Nielsen-Englyst, C. (2025). *Grønne fællesskaber og carbon fodaftryk* [*Green communities and carbon footprint*]. SAMSKAB and Copenhagen University. https://doi.org/10.13140/RG.2.2.22464.60165
Grinde, B., Nes, R.B., MacDonald, I.F., & Wilson, D.S. (2018). Quality of life in intentional communities. *Social Indicators Research*, 137, 625–640.
Grum, B., & Grum, D.K. (2020). Concepts of social sustainability based on social infrastructure and quality of life. *Facilities*, 38(11/12), 783–800.
Hamdi, N. (2004). *Small change*. Routledge.
Hoff, J., & Gausset, Q. (Eds.) (2016). *Community governance and citizen-driven initiatives in climate change mitigation*. Earthscan.
Hoff, J., Gausset, Q., & Lex, S. (Eds.) (2020a). *The role of non-state actors in the green transition: Building a sustainable future*. Routledge.
Hoff, J., Gausset, Q., & Lex, S.W. (2020b). Introduction. In J. Hoff, Q. Gausset, & S. W. Lex (Eds.), *The role of non-state actors in the green transition. Building a sustainable future* (pp. 1–13). Routledge.
Jackson, T. (2005). Live better by consuming less? Is there a 'double dividend' in sustainable consumption? *Journal of Industrial Ecology*, 9, 19–36. https://doi.org/10.1162/1088198054084734
Jarvis, H. (2015). Towards a deeper understanding of the social architecture of cohousing: Evidence from the UK, USA and Australia. *Urban Research & Practice*, 8, 93–105.
Lallement, M. (2024). *A desire for equality: Living and working in concrete utopian communities*. Bristol University Press.
Lam, D.P., Martín-López, B., Wiek, A., Bennett, E.M., Frantzeskaki, N., Horcea-Milcu, A.I., & Lang, D.J. (2020). Scaling the impact of sustainability initiatives: A typology of amplification processes. *Urban Transformations*, 2(1), 3.
Lane, R.E. (1993). Does money buy happiness? *The Public Interest*, 113, 56–64.

Lee, D.R. (2005). Who says money cannot buy happiness? *The Independent Review*, 10(3), 385–400. www.independent.org/pdf/tir/tir_10_3_05_lee.pdf
Levitas, R. (2008). Pragmatism, utopia and anti-utopia. *Critical Horizons*, 9(1), 42–59. https://doi.org/10.1558/crit.v9i1.42
Levitas, R. (2013). Some varieties of utopian method. *Irish Journal of Sociology*, 21(2), 41–50.
Lietaert, M. (2010). Cohousing's relevance to degrowth theories. *Journal of Cleaner Production*, 18, 576–580.
Lockyer, J. (2017). Community, commons, and degrowth at Dancing Rabbit Ecovillage. *Journal of Political Ecology*, 24, 519–542.
Lockyer, J., & Veteto, J.R. (Eds.) (2013). *Environmental anthropology engaging ecotopia: Bioregionalism, permaculture, and ecovillages*. Berghahn.
Lubell, M. (2002). Environmental activism as collective action. *Environment and Behavior*, 34(4), 431–454. https://doi.org/10.1177/00116502034004002
Manzella, J.C. (2010). *Common purse, uncommon future: The long, strange trip of communes and other intentional communities*. Bloomsbury.
Moore, M.L., Riddell, D., & Vocisano, D. (2015). Scaling out, scaling up, scaling deep: Strategies of non-profits in advancing systemic social innovation. *Journal of Corporate Citizenship*, 58, 67–84.
Mulder, K., Costanza, R., & Erickson, J. (2005). The contribution of built, human, social and natural capital to quality of life in intentional and unintentional communities. *Ecological Economics*, 59, 13–23.
Nelson, A., & Chatterton, P. (2022). Dwelling beyond growth: Negotiating the state, mutualism and commons. In F. Savini, A. Ferreira, & K. von Schönfeld (Eds.), *Post-growth planning: Cities beyond the market economy* (pp. 49–62). Routledge.
Nguyen, T. (2019). *The production of community in Munksøgård: Community, values and infrastructures*. Department of Anthology, University College London.
Oishi, S., Kesebir, S., & Diener, E. (2011). Income inequality and happiness. *Psychological Science*, 22, 1095–1100. doi: 10.1177/0956797611417262.
Omann, I., Kammerlander, M., Jäger, J., Bisaro, A., & Tàbara, J.D. (2020). Assessing opportunities for scaling out, up and deep of win-win solutions for a sustainable world. *Climatic Change*, 160(4), 753–767.
Ortiz-Ospina, E., & Roser, M. (2020). *Loneliness and social connections*. Published online at OurWorldInData.org. https://ourworldindata.org/social-connections-and-loneliness
Palgi, M., & Getz, S. (Eds.) (2025). *Conflicts and conflict management in intentional communities*. Walter de Gruyter.
Pellizzoni, L. (2020). Prefiguration, subtraction and emancipation. *Social Movement Studies*, 20(3), 364–379. https://doi.org/10.1080/14742837.2020.1752169
Pickerill, J. (2025). *Eco-communities: Surviving well together*. Bloomsbury.
Rootes, C. (2003). The transformation of environmental activism: An introduction. In C. Rootes (Ed.), *Environmental protest in western europe* (pp. 1–19). Oxford University Press.
Sahlins, M. (1972). *Stone Age economics*. Aldine Atherton.
Sanford, A.W. (2017). *Living sustainably: What intentional communities can teach us about democracy, simplicity, and nonviolence*. University Press of Kentucky.
Sargent, L.T. (2010). *Utopianism: A very short introduction*. Oxford University Press.

Sargisson, L. (2012). 9 Green Utopias of Self and Other. In B. Goodwin (Ed.), *The philosophy of utopia* (pp. 140–156). Routledge.
Satici, S.A., Uysal, R., & Deniz, M.E. (2016). Linking social connectedness to loneliness: The mediating role of subjective happiness. *Personality and Individual Differences*, 97, 306–310. https://doi.org/10.1016/j.paid.2015.11.035.
Schehr, R.C. (1997). *Dynamic utopia: Establishing intentional communities as a new social movement*. Bergin & Garvey.
Shenker, B. (2011). *Intentional communities: Ideology and alienation in communal societies*. Routledge.
Sherry, J. (2014). *Community supported sustainability: How ecovillages model more sustainable community*. [Doctoral dissertation, The State University of New Jersey]. http://dx.doi.org/10.7282/T3959FVM
Toft, M., Gausset, Q., Hoff, J.V., & Lex, S.W. (2021). *Together we green: Measuring the community effect on environmental behaviour*. Copenhagen University. [Unpublished manuscript].
Törnberg, A. (2021). Prefigurative politics and social change: A typology drawing on transition studies. *Distinktion [Distinction]: Journal of Social Theory*, 22(1), 83–107. https://doi.org/10.1080/1600910X.2020.1856161
Vita, G., Ivanova, D., Dumitru, A., García-Mira, R., Carrus, G., Stadler, K., ..., Hertwich, E.G. (2020). Happier with less? Members of European environmental grassroots initiatives reconcile lower carbon footprints with higher life satisfaction and income increases. *Energy Research & Social Science*, 60, 1–11.
Wang, H., Cheng, Z., & Smyth, R. (2017). Consumption and happiness. *The Journal of Development Studies*, 55(1), 120–136. https://doi.org/10.1080/00220388.2017.1371294
Wapner, P. (1996). *Environmental activism and world civic politics*. State University of New York Press.
Wilkinson, R., & Pickett, K. (2009). *The spirit level: Why equality is better for everyone*. Penguin.
Wright, E.O. (2020). *Envisioning real utopias*. Verso Books.

1

PEOPLE-POWERED GREEN PLACEMAKING IN THE UNITED KINGDOM

Multi-level strategies and a theory of change

Helen Jarvis

Introduction

This chapter discusses the significance of a theory of change to achieving the green transition from a sustainable degrowth perspective. A theory of change is a dialogue-based process that describes the sequence of steps a group of people expects to follow to achieve a desired outcome. It helps the group conceptualise how to make change happen in each situation, usually by emphasising interpersonal relationships of collective action and relational power. Edward Chambers (2017: 12–13) observes that 'building relationships for action is ultimately more powerful, long-lasting, and satisfying than accomplishing a series of tasks'. He contrasts the Latin origins of the word 'relationship', which means 'to relate what happened', with the Latin origins of the word 'task': 'taxa', which means to 'rate' or 'value' something. Focusing on a theory of change can help explain why green communities remain niche, or specialised, in scope and scale. These communities comprise close-knit interpersonal relationships that are torn between learning and practising how to live and work well together as a group while also building their collective power to create change.

For more than two decades, my research has critically engaged with groups that organise themselves in opposition to dominant cultures prioritising individualism by focusing on the hyper-local micro-social phenomena entailed in groups that choose to create and sustain socially and ecologically just systems prioritising interdependence. In this context, a theory of change can be perceived as a fashionable mantra added to a long list of goals. If community-led change was easy and effective, we would see democracy flourishing. As it is, we are experiencing a decrease in democratic societies

DOI: 10.4324/9781003769064-2

around the world. However, it is important to distinguish the persuasive logic and relational power of a theory of change (if-then-because-why?) on a continuum, from popular tactics of protest (petitions and boycotts) and practical services (repair cafés and seed swaps) to efforts to prefigure a desired change by living 'as if' it has already been achieved (Schiller-Merkens 2024), and system change.

The umbrella term 'intentional community' (IC) is applied to groups that co-create and occupy clustered homes based on shared values and a desire to live more lightly on the planet by owning and consuming less 'stuff' (such as cohousing and eco-villages). Both cohousing and eco-villages are living examples of the commitment to living more lightly.

Whole books are devoted to describing community efforts to distinguish sub-categories of 'intentionality' practiced by different groups and their living, working, and income-pooling arrangements, for example. The wide-ranging scope of the idealised and pragmatic IC is *not* the focus of interest here. Instead, this chapter discusses insights from IC research and applies them to dispersed placemaking, i.e., people working together to improve and transform a place they inhabit or care about. The aim is to trace similar attributes of interpersonal connection, such as trust and purpose with preparing food and eating meals together, as the 'glue' of relational power that drives collective action and effective change (Jarvis 2011, 2015, 2019). By consciously interacting in relational togetherness, citizens from different walks of life build alliances and organise change for a common purpose.

Popular expectations that state actors consult with local citizens about investment in a local area rarely correspond with meaningful expressions of democracy and fairness. For example, some situations of austerity urbanism in the United Kingdom (UK), where government authorities devolve power and decisions to local people, ultimately reinforce structural inequalities by engaging community groups in limited, token consultation (Sara et al. 2020: 494). By contrast, this chapter draws on the vocabulary of degrowth, i.e., a movement that critiques the pursuit of economic growth at all costs and instead prioritises social and ecological well-being through a reduction and redistribution of production and consumption, to critically engage with the transformative power of groups. A degrowth perspective challenges the prevailing bias toward consumption and the organising principles of what is called 'consumer citizenship' in the social and commercial life of traditional, neoliberal market economies across diverse local cultural settings (D'Alisa et al. 2015).

There is widespread concern in the UK that many of its towns and cities have a 'torn social fabric', that many communities exhibit fear, grievance, and lack of trust in authority (Onwards 2024). This pattern follows years of austerity and the damaging effects of individualism, inequality, and people who live isolated from others. The Onwards thinktank (2024) advanced a social fabric index to assess the density and strength of social assets and interpersonal relationships in different places. The index also measures goods, services, and intangible qualities of civic infrastructure, such as trust in government, social prescribing (health services connecting people to non-state activities and groups in community), volunteering, climate change adaptations, and the existence of green spaces.

This social fabric index largely restates a 'levelling-up agenda' that calls for investment in new physical places and community organisations (libraries, transportation hubs, churches, and parks). The prevailing wisdom implies a direct correlation between physical spaces of social infrastructure and healthy community engagement (Latham & Layton 2022). Yet, the index says little about the capacity for people to organise collectively or pursue a shared vision with purpose. Significantly, Onwards (2024: 68) accept that the physical infrastructure of a place should not be prioritised to the exclusion of social factors such as purposeful relationships and new and deeper forms of democracy. However, little is known or taught about all that is 'secret' and unrecognised in human-environmental connectedness, whether living together in the clustered homes of an IC or in dispersed green communities.

Most existing urban, community, and development studies neglect informal and intangible aspects of placemaking. They focus on individual actors rather than the collectively agreed-upon 'we' of interpersonal connection: places are said to be 'made' according to 'top-down' planners and marketing initiatives (Richards 2021). Places are then shaped from the bottom up through the attitudes, behaviour, and choice (ABC) of consumer citizens (Donaldson & João 2020; Shove 2010). Niche grassroots innovations further shape placemaking by introducing and expanding principles of democratic control by members and residents, alongside stakeholders from the local community (Lang et al. 2020). Examples include community groups forming housing and land trusts and taking ownership of communal spaces such as parks and shops. This is illustrated by loyal customers who took over a family-run hardware supplies store in Kings Heath in Birmingham, UK to prevent its being sold to a developer when the owners retired (Sara et al. 2020; https://yorksuppliesco-op.co.uk/). These are rare situations, where local residents are sufficiently well-organised to amplify their collective power in pursuit of a common goal, in this case continuing the legacy of a 'kind place' to shop that aligns with a sustainable degrowth ethos (Power to Change 2025).

In practice, placemaking governance is messier and multi-layered. Civic leaders and community groups are motivated by diverse 'civic imaginations', or visions of their future (Richards 2021). Individual activists and collective endeavours are always also embedded in complex multi-level systems. This is important to acknowledge for two reasons. On the one hand, community engagement and a sense of place and belonging constitute intersecting interdependencies within and between the scope of household, neighbourhood, workplace, social purpose, cultural practice, and everyday lived experience (Morgan 2020). These relational practices largely determine spheres of influence, why and where people feel compelled to act on issues that concern them, and how people organise with others to amplify their power and agency. On the other hand, we are all bound up in climate-related risks and anxieties that connect the local to global structures and institutions, *here-and-now* and over the long term. Arguably, sustainable degrowth transitions call for top-down as well as bottom-up, multi-level innovations, reforms, and systems change. In short, this messiness describes multiple levels, identifications, and dimensions of space, time, and coordination.

I observe elsewhere in this chapter the irony that what is 'secret' in official accounts of urban placemaking is in reality part of everyday life (Jarvis et al. 2001; Jarvis 2018). The intricacies of forming and sustaining interpersonal relationships of togetherness are not strictly secret but directly experienced and consciously practiced by those who are immersed in a community of practice or project, i.e., negotiating how to get a project off the ground. Yet this realm of connection *is* 'secret' in that it is unrecognised in formal, counted, measurable definitions and discussions of urbanism and in terms of the policy responses to what are identified 'top down' (in formal ways) as deficits and exclusions in community cohesion and place attachment (Jarvis et al. 2001; Jarvis 2019). In response, I counter this neglect by consciously highlighting the multi-level social architectures and relational 'glue' of hyper-local micro-activism and draw insights from the different ways that activists, practitioners, and community groups organise together in broad alliances to develop or retrofit sustainable degrowth placemaking in the UK (Civic Square 2023).

This chapter is structured in three parts. First, a multi-level conceptual framework is introduced to articulate the messiness of hyper-local community engagement. This framework advances a process-oriented understanding of grassroots innovation and sustainability transitions and builds on previously published approaches applied to housing and domestic energy systems (Geels & Schot 2007; Lang et al. 2020). I use this framing to shift the emphasis from the quantity of social interaction (how much people interact in public space) (Gehl & Svarre 2013: 50) to understanding *strategies of purpose*, which is fundamental to a theory of change.

Second, practical examples of hyper-local activism are introduced to illustrate how a continuum of practice and motivation coexist within and between multiple scales of individual and collective strategies of purpose and a theory of change. Finally, the chapter concludes with an invitation to activist placemakers to combine their immediate goal to change their own practices and inhabit low carbon neighbourhood projects with an enduring political commitment to advocate for national and global systems change. As Jens Hoff et al. (2020) observe, new forms of green group *practivism* are replacing older forms of environmental protest and non-violent direct action. Anyone wanting to achieve sustainable degrowth should switch from movements of protest to plans of action and root these actions in pragmatic theories and methods of change. The chapter's conclusions align with seminal texts by Deepak Bhargava and Stephanie Luce (2023) and Erik Olin Wright (2013) on theories of change that determine effective and pragmatic collective action for the here and now.

Interdependent levels and communities of practice

A multi-level approach that is temporal as well as spatial reminds us that we cannot afford to wait until conditions are ripe for an ideal or united response to all that is wrong in 'the world as it is'. Cooperation rarely functions through complete agreement or unity of purpose. Solidarity-based networks must negotiate with a broad network of stakeholders from the local community as well as state and market interests. This contrasts with the often romanticised, positive, and parochial connotations of 'place', 'community' and 'solidarity'. There are good reasons, therefore, to campaign on issues and places in the 'here-and-now' in local situations of collective action, where individual power and agency are amplified to be most effective and influential.

However, while it is largely true that 'the most powerful form of activism is just the way we live' (Daly 2018), especially in the context of ICs prefiguring a green-sharing economy, lifestyle practices alone exert little in the way of effective change, and activists occupy multiple levels and communities of practice interdependently. According to Etienne Wenger (2000), communities of practice comprise three transformative dynamics that are the basis of a theory of change. These dynamics hold people accountable by a sense of collectively agreed joint enterprise, they establish norms and relationships of mutuality that reflect interactions of mutual engagement, and they produce a shared repertoire of communal resources—language, routines, sensibilities, artefacts, tools, stories, styles, and the like (Wenger 2000: 229).

Frank Geels and others define the multi-level approach by distinguishing three interdependent processes that articulate how diverse stakeholders and communities of practice simultaneously engage in democratic and sustainable placemaking: niche, regime, and landscape (Geels & Schot 2007). Richard Lang et al. (2020) apply this analysis to recent developments in community-led housing, a grassroots-based housing niche in England.

The *niche* is where we find ICs and community-led placemaking initiatives that serve as laboratories for testing and demonstrating new ideologies and social structures in practical ways, such as building car-free clustered housing with shared growing spaces.

The *regime* represents the combined structural features, rules, and institutions that reproduce individualistic cultural attitudes and consumer choices, as well as the welfare initiatives and intended 'safety nets' that variously restrict and redistribute access to private and public goods and services. For example, from a historical-comparative analysis of state interventions in housing and welfare, the UK is a regime of high home-ownership rooted in pro-home ownership policies also associated with declining affordability and social inequality (Forrest et al. 2021).

The *landscape* represents the slow changing domain of 'world views and paradigms, macro-economy and material infrastructure, as well as the natural environment and demographics' (Bergman et al. 2007: 3). Both regime and landscape systems are difficult to change because they are entrenched and stabilised by various lock-in mechanisms, and the niche typically lacks power and scale (Geels & Schot 2007).

Christer Sanne argues that the promotion of both private home and car ownership constitute lock-in mechanisms that presuppose heavy capital and fiscal investments (motorways and car parks). These mechanisms are in contrast to niche green housing and transport projects, such as environments designed to promote walking and cycling, where state and market investment is minimal (Sanne 2002: 282). This niche represents small groups of green activists and methods of community organising that are typically at odds with the dominant regime. The significance of the niche is that it harbours the potential for testing and demonstrating *practivism* (Hoff et al. 2020), which embeds innovation to make it more effective. Community-level social learning is also an important and often overlooked function of hyper-local activism that I return to below.

Arguably, the 'demonstration' effect of degrowth placemaking is necessary but insufficient for a comprehensive and just green transition. Demonstration is the idea that activist groups are inspired by 'seeing to believe' that a better future is possible by visiting and learning about existing niche innovations (Nelson 2018: 186). The concept of a breakthrough effect in the transition to cut carbon emissions to net-zero (a 20% adoption of renewable technology) is usually applied in traditional communities to the roll-out of 'hard' technologies such as electric vehicles and renewable home energy sources rather than to 'soft' cultural practices such as communal living arrangements, clothing swaps, and neighbourhood gardening groups that are prioritised in green communities (Bergman et al. 2007).

However, it is useful to conceptualise equivalent diffusion of lifestyle practices on a continuum that identifies a series of niche innovators: pioneers, early adopters, early majority, late majority, and laggards. These innovators are shown in Figure 1.1: they are mapped across the interdependent analytical levels of niche, regime, and landscape described above. It is widely observed that niche low-carbon projects and lifestyle practices rarely achieve a breakthrough effect, i.e., they rarely become mature eco-communities, even if some are popular and successful 'demonstration' projects. This is because

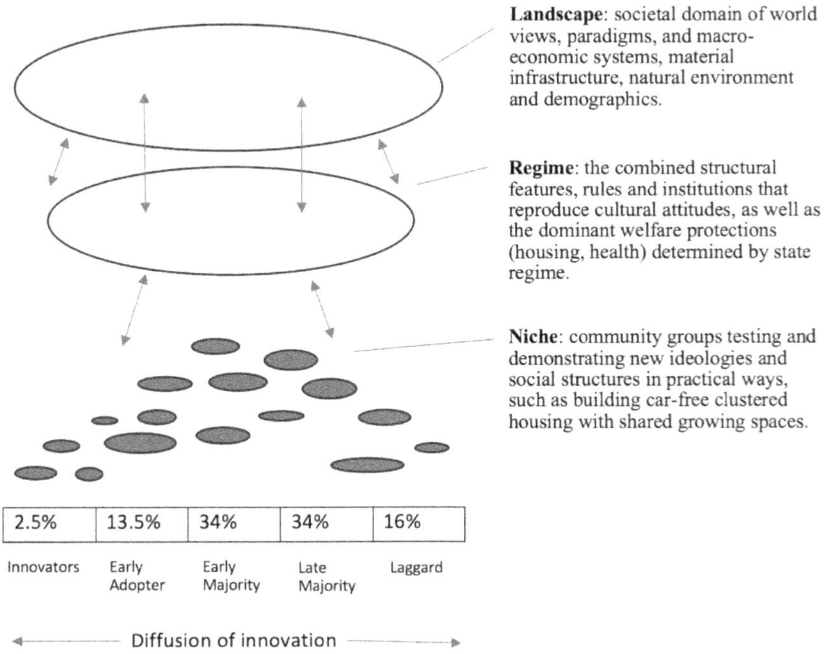

FIGURE 1.1 Multi-level diffusion of innovation showing three levels: landscape, regime, and niche

individual pioneers and early adopters can become 'stuck' at an early stage of becoming a true green community, or they experience 'burnout' before it is possible for that niche to become sufficiently empowered to transform a local or national regime (Lang et al. 2020: 54; Boyer 2015). Arguably, a theory of change for the IC is not limited to empowering niche innovations 'at scale'; there are arguments for scaling *out*, rather than scaling 'up' (Heywood 2016; Constantino et al. 2022). Rather, the theory calls for groups to develop their interpersonal skills of relational power in order to work in partnership with sympathetic practitioners and to avoid the group attrition of and the collapse of an entire project (Sanford 2013).

Much of what we read and hear about climate change and social inequality is disempowering. Feeling powerless is a moral hazard: the knowledge that we are doubtless part of the problem and not bearing the worst extremes of heating, fire, and flooding can contribute to our feeling overwhelmed by climate anxiety, guilt, shame, and hopelessness (Ray 2020: 115). How we deal with feeling anxious, whether on our own or jointly with others, is a key element in the distinctions between older and newer forms of activism (Chatterton 2023: 110). Arguably, barriers to niche empowerment are as frequently social and organisational as they are material or technological.

Individuals who are drawn to activist groups typically reflect on their own and other people's personal politics, social networks, and social responsibility in the process of aligning their actions with a specific purpose. It can feel overwhelming to lose oneself in a bigger sphere of influence. High rates of activist burnout among individual green pioneers (shown here as around 2.5% of the population in Figure 1.1) can be explained by social and physical isolation as well as underdeveloped skills of cooperation.

Few of us are well-equipped to align self-interest with collective benefit because conventional education tends to advance instrumental rather than compassionate or reflective learning. According to Richard Sennett (2012: 29), 'people's capacities for cooperation are far greater and more complex than institutions allow them to be'. Yet the skills and capacities required to form and maintain inclusive and cohesive communities often flounder without the support of the dominant regime. For instance, the following chapter by Ana Margarida Esteves concludes that social inclusiveness is a prerequisite for the scalability of green communities.

When I apply a multi-level analysis, I find that a scalable and just green transition is possible when grassroots groups join with other stakeholders such as social landlords or private and public sector practitioners. I consider how this alignment of activist orientations might arise in the following section by using desk research and ethnographic studies that I've conducted in recent years. Five hyper-local placemaking initiatives are selected from UK cities to illustrate how green groups and communities can retrofit a sense of place and purpose within existing social and environmental systems.

Geographies of proximity

Empirical observations presented in overview in Box 1.1 are illustrative, not representative. The aim is to make practical sense of the largely theoretical discussion above. Examples of placemaking were selected to illustrate interdependent social, temporal, and spatial 'architectures of cooperation'. These examples demonstrate a common pattern and process of enduring relationships and collective endeavour that typically involve collective management (if not ownership) of shared community space and coordinated conviviality (for example, the community preparing and enjoying meals and celebrations together on a regular basis). These autonomous groups differ from service-delivery and advocacy non-profits in the common sense that they are oriented toward implementing practical and effective improvements to the quality of place and lives of members.

All five examples in Box 1.1 involve physical spaces and places, whether permanent or temporary, which function as 'social spaces' for convivial and potentially transformative relationships. They all have strategies for collective governance and conflict resolution. The point is that 'real' social architectures arise from the collectively negotiated habits of 'making' these places. These examples illustrate the interdependency of space, time, and coordination that empower grassroots placemaking, i.e., social spaces mobilise different forms of socially productive ownership and management of community assets.

Groups are observed to consciously interact in shared spaces of togetherness by developing social architectures of cooperation, by coordinating social and physical spaces in real-time, and through the evolution of group dynamics (Jarvis 2011). Underpinning much of my previously published IC ethnographic research is the role of proximity to conviviality beyond the distinction of real-time 'in person' from virtual gatherings: residents live together under one roof or many, combining access to shared-with-private spaces. IC groups escape the dominant norms of individualism and materialism by adopting a consciously devised and usually well thought-out social and cultural alternative to single-family dwelling. In the pursuit of these goals, they share significant aspects of their lives together (Jarvis 2015). The IC is largely premised on the idea of replacing individualised with collectivised responsibility for transitions to sustainable degrowth (Jarvis 2019).

By default, the dominant regime relies on individual motivation and agency to 'reduce' (wasteful consumption), 'reuse' (or re-purpose) goods where possible (avoiding single-use plastics), and 'recycle' (goods and waste), forming a circular economy. Conventional home and neighbourhood designs make it difficult to forge friendships and develop trust, mutuality, solidarity, and togetherness. The individual consumer has far less power to resist or oppose the dominant regime than a coordinated group. As Ma'ikwe Ludwig (2017: 13) observes, 'every (niche) intentional community is experimental to some extent... attempting something cooperative within the context of a highly competitive culture'. For example, it is common for ICs to organise a green sharing alternative such as reciprocal childcare, electric car clubs, mutual aid groups and the like (Jarvis 2011).

The focus of ICs on shared property suggests that it takes more than trivial interactions and encounters (e.g., routinely cooking and eating together) to establish and maintain a cohesive community. For groups collaborating on a project where the end goal is to share living spaces and manage their domestic lives together, the personal stakes are high. This point is made well by Kate Nichols in *Communities* magazine when she reflects on what it was like to get a cohousing community off the ground in the state of Washington, US, in 1999:

> (It) was one of the most exhilarating and exhausting projects I've ever participated in, When I look back at the success of the project, much of it (finding the site and financing construction) was serendipitous... (but) under the pressure of getting the project built, we ignored some of the

> problems within the community... After we moved in, it became apparent that in rushing to complete the buildings, we had failed to form a solid community, one where we cared about each other. Although we had some training in consensus, members weren't always respectful to each other. The unresolved tensions and burnout were such that five out of the original six founding members moved out within the first few years.
>
> *(Nichols 2024: 28)*

Once groups have formed and inhabited a shared property, communitarians typically regard themselves as a continuing group, separate from and in many ways better than, the society from which they have emerged. However, it is not uncommon for members to become disillusioned or to observe that their group was better equipped to manage the construction of their homes than to nurture and hold their group together emotionally, and take time to build enduring trust and care for one another in order to resolve tensions and conflict as they arise. By comparing ethnographic observations for co-located and geographically dispersed communities of practice, it is evident that all groups face similar challenges when reconciling their shared vision, project, or 'task', such as forming a cooperative to crowdfund the purchase of a community asset.

Groups also recognise the need to invest in relationships and the residents' collective wellbeing that will bind their community together over the long-run. The 'task' function refers to the work and activities that the group undertakes to fulfil their purpose (to share resources, to live lightly on the planet). The 'relationship' functions require advanced social learning, which can sometimes exhaust the capacity of small or busy groups (Ochre 2013; Jarvis 2025). Significantly, the latter is essential if collective action is to be effective as a strategy of purpose and theory of change.

This analysis echoes the distinction that Sennett makes between 'hard' and 'soft' 'architectures of cooperation' in cities and urban neighbourhoods as well as in the blurred edges of these social scales. Sennett (2012a) distinguishes between the 'hard' built environment of the city regime and the 'soft' architectures that comprise the way of life and conviviality of neighbourhood attachments. The language of hard and soft and blurred is evocative, in contrast with the priority usually given to physical social infrastructures in the UK levelling-up agenda. Such language describes how activist groups acquire and develop the skills of cooperation and conflict resolution by interacting relationally. Tacit skills that are rooted in lived experience permeate group formation and both fragile and enduring relationships of mutual trust and care (Ochre 2013). There is inevitable tension between the collective action required to maintain group relationships and shared property and a theory and strategy of wider placemaking. This is further explored in the following analysis.

BOX 1.1 EVIDENCE OF 'SOFT' PLACEMAKING: COMPARISON OF FIVE UK GREEN COMMUNITY CASE STUDIES

(i) Dwellbeing Shieldfield, Newcastle upon Tyne, UK

Dwellbeing is a community cooperative since 2017 (and constituted as a Community Benefit Company since 2021 (www.dwellbeingshieldfield.org.uk/how-we-work)) in the East of Newcastle upon Tyne, UK, an area of deprivation and regeneration. Members of the cooperative share life together, support one another, and work to transform the Shieldfield area into a good place to live.

The cooperative uses a participative approach to address the structural issues that have beset the neighbourhood in recent years. The theory of change that the cooperative illustrates is driven by 'stewards' (trustees), worker members (staff team), and other members, all of whom live and work in Shieldfield and are collectively organised. Member-led weekly gatherings (through walks, trips, talks, discussions, and food-sharing) provide the time and place for neighbours to support each other, build friendships, and provide space for skill-sharing and education in the rhythms of ordinary life. The programme involves urban growing, public realm improvements, and building community knowledge and engagement around issues of food security and youth development (www.rethinkingpoverty.org.uk/local-initiatives/dwellbeing-shieldfield/).

(ii) Civic Square, Ladywood, Birmingham, UK

Civic Square (constituted in 2014 as a Community Interest Company) demonstrates neighbourhood-scale civic infrastructure for social, ecological, economic, and climate transition together with many people and partners. The project includes several programmes, including 'Neighbourhood Transitions', a street-based programme that focuses on people-powered social and built net-zero transitions. This alliance has co-authored a toolkit, 'Neighbourhood Public Square', a demonstrator project that is part of the wider ecology of transformation.

The Retrofit Reimagined 2023 festival season encouraged a network of experts and community projects to come together to support the scale of transformation enabled by a people-powered retrofit manifesto and knowledge exchange. A compelling 'argument for hyperlocal integrity ethics in retrofit construction is made on the strengths of trust-building conversations and pooling local skills', rather than out-of-place outside consultant assumptions. Keeping information, trust, contractors, and passion hyper-local is important: it is the 'glue' that holds relationships together. Quoting Florence Collier Humblebee on the use of hemp insulation, 'We need people to advocate for retrofit of their streets and communities: "neighbours telling neighbours" ' (https://civicsquare.cc/).

(iii) Citizens House, Lewisham, UK

This community-led housing initiative grew from a broad-based community organising (BBCO) campaign led by Lewisham Citizens (a chapter of Citizens UK) that listened to local concerns about the lack of affordable homes and organised to find a solution as a Community Land Trust (CLT). The partnership of London CLT with Lewisham Citizens combined methods for and training in 'speaking truth to power' (building relationships with the elected Mayor of Greater London and Lambeth Council leaders in land use planning) with skills and capacity in community housing and sustainable placemaking (Source: Personal communication with Hannah Gretton, Lewisham Citizens, September 2024).

This case helps distinguish between grassroots intentional communities (led by a core group who wish to house themselves) and a scheme driven by community benefit groups (London CLT and Lewisham Citizens, involving a limited number of prospective residents in the plans). Citizens House is fully occupied and successful in material terms but aspects of building a community such as relationships of trust and shaping sustainable placemaking through social learning were neglected due to the arms-length scope of BBCO. Design workshops were organised with residents from the wider neighbourhood rather than prospective residents of London CLT. The lack of a common space and shared meeting times limits the scope of residents organising a shared social life and mutual aid, but as an institutional member of Lewisham Citizens and part of London CLT, Citizens House residents are part of a wider alliance (https://www.londonclt.org/citizens-house).

(iv) Rural Urban Synthesis Society (RUSS), London, UK

RUSS is a members-led Community Land Trust (CLT) based in southeast London founded in 2009) with members occupying 36 permanently affordable eco-homes, the largest project of its kind ever undertaken in London from 2023. The completed project comprises two apartment blocks surrounding a car-free community space with lots of natural planting. The homes are on several levels and were built using modern methods and sustainable materials. Each home was tailored to the resident member, who helped design the scheme. Member-residents are assured permanently affordable homes (rented from the CLT) with very low energy costs. RUSS is a member of Lewisham Citizens (see case iii above).

RUSS co-founder Kareem Dayes was inspired by pioneering timber-framed designs and the lived experience of shared housing from the 1980s. 'It started off really because of the photovoltaics and the whole energy side of things, thinking, ok, so a house can generate its own energy. This idea that you can go from consumers to producers (of energy)... the idea of micro-generation, decentralisation, off-the-grid, really captured my imagination... building this as a CLT, (we found a model) that banks would lend to, the council would give land for, and that they would grant planning permission for. So it was turning that vision into a community-led project that could be built and financed' (Kareem Dayes quote in YouTube video transcript: From Walter Segal to RUSS, 23 October 2018 (https://www.youtube.com/watch?v=qba-OlrqcMY).

(v) Mushroom Works, Newcastle upon Tyne, UK

The artistic collective Mushroom Works was constituted in 2006 as a Community Interest Company (CIC) to regenerate and occupy a 19th-century former Wesleyan chapel to accommodate 12 studios for 16 creative practitioners plus gallery space in a post-industrial 'creative quarter' area of regeneration on Tyneside. This is a non-hierarchical Community Interest Company with the scope to cultivate deep relationships of social purpose and trust and a common concern for democratic and sustainable placemaking. Tenant artists organise themselves according to tacit expectations of community spirit and social purpose, where the whole is greater than the sum of its parts. This is witnessed in rituals of shared space, social time, and mutual support: 'Three times a day the tea bell is rung and the working collective congregate in the kitchen to share thoughts and ideas and to take a break from their creative toil. They think of themselves as an extended family, similar, in a way, to the fellowship (of the former chapel) that gathered here twice daily 150 years ago' (https://mushroomworks.com/).

Hyper-local micro-activism

The five examples of hyper-local placemaking presented in Box 1.1 illustrate common social architectures of interpersonal connection. The data combines desk research with previously published IC research, including participant observations from my role in the leadership group of Tyne and Wear Citizens since 2018 and field site visits with undergraduate students in a programme of community-engaged learning (CUK 2025; Jarvis 2023). (Tyne and Wear Citizens is a chapter of the national Citizens UK NGO.) Interpersonal group phenomena are interpreted and communicated using a composite vignette method (Jarvis & Bonnett 2013: 2355). This method conveys succinctly the rich complexity and the intersection of individual biography and myriad group-work situations while masking individual subject identities.

Examples of hyper-local placemaking are selected to illustrate interdependent social, temporal, and spatial 'architectures of cooperation', in order to reveal the 'secret' attributes of collectively negotiated 'we' thinking that are often overlooked in official accounts. Members of the placemaking group Dwellbeing, for example, find common ground in their lived experience of a social rented housing estate in a low-income neighbourhood that has been historically neglected by mainstream economic development. Members and residents participate in mutual networks of reciprocity but they do little to coordinate energy conservation or affordability within individual homes and households. Civic Square is a broad-based alliance of green activist groups that focus on hyper-local projects distributed across several target neighbourhoods in the city of Birmingham. There are several resident cooperatives within the alliance, but members draw significantly from practitioners rather than resident communities, without living together, which is similar to the artist collective at Mushroom Works on Tyneside.

Citizens House and Rural Urban Synthesis Society (RUSS) are both community land trusts (CLTs) that provide permanently affordable homes in London. Each has followed a discretely different approach to group and project development. Broadly speaking, Citizens House was developed for community benefit, whereas RUSS was planned and developed by and for a core group of pioneer members as an IC. Although Citizens House and RUSS propose ostensibly different geographies of proximity, closer scrutiny of lived experience in these projects indicate similar pressures to balance group task requirements with the relational work of nurturing and holding each community together. This balancing of individual and collective responsibilities resonates with how we deal with feeling anxious about climate change and social inequality, as described above, whether on our own or jointly with others.

RUSS is the closest match to previous research on inhabited ICs. For more than a decade the group gathered regularly, functioning as a close community long before moving in to their clustered homes together. Founding member Kareem Dayes drew inspiration from witnessing pioneering self-build and affordable self-managed housing cooperatives in the 1980s, looking for ways to reimagine affordable green community housing today. The challenge remains: how to 'scale up' this individual project (which took 15 years to realise) and use it to inspire and amplify the early adoption stage of an empowered niche. All five examples demonstrate a common pattern of enduring relationships and collective endeavour, including thoughtful group processes and collective management (if not ownership) of shared community space. Each group pursues its own 'call for change' within which there is 'an ethics that privileges emotions, relationships, and experiences over impartial and rational moral reasoning' (Schalk et al. 2017: 447). Paradoxically, whether or not they ever intended to live or work together as a close community, from an early stage of group formation and strategising, these different groups all reproduce the 'togetherness' typically associated with living communally.

They do this through shared meals, seasonal celebrations, shared projects, and educational retreats.

Ethnographic studies reveal that working together and sharing meals are primary bonding experiences that are not unique to the shared spaces and social times of the established IC. This reflects the widely accepted understanding that shared meals that neighbours prepare and eat together are the glue that binds and endows meaning to community relations (Blank 2001; Christian 2003). The social significance of the nurturing aspects of communality, even when living apart, represents a space of possibility for 'we' thinking. The point is that these nurturing practices and relationships are *not* transactional or reduced to individuals protesting or providing a service (Jarvis 2011, 2015, 2019). This is highlighted in the case of the studio project Mushroom Works, where artists who identify as sole traders gather together for refreshments each day and cherish this time together as a source of solidarity and a way of preventing emotional burnout. As Litfin (2014: 128) observes, the simple act of working together is a primary bonding experience.

These autonomous groups differ from service-delivery and advocacy non-profits in the fundamental sense that members are similarly motivated to pursue a common goal that unites them and sustains their commitment to a theory and method of achieving change as an alternative and better future. All of these examples involve physical spaces and places, whether permanent or temporary, which function as social spaces for shared time and a rhythm of habits that are convivial and coordinated by agreed rules and responsibilities. These communities of practice exhibit all three dimensions of dynamic and transformative potential identified by Wenger (2000) as the basis of a theory of change, even if in subtle and 'secret' ways. The point is that the real social architectures are the collectively negotiated habits that imbue these places with purpose and meaning (conviviality, consensus, conflict resolution, etc.). While it matters that interactions and encounters are to a large extent real, not virtual, it can be interpreted from Box 1.1 that physical spaces and real-time synchronisations are not the only coordinates of the glue that matter. The analysis highlights multi-level relationships of care, compassion, and the time that it takes to build enduring relationships of trust: social spaces mobilise different forms of socially productive ownership and management of community assets (Bunce 2020) in order to amplify the relational power of grassroots placemaking.

A consistent feature of group maintenance and community organising practices in Box 1.1 involves the scheduling and coordination of team rotas, or rosters, for seasonal rituals and celebrations that are intended to remind the group of its purpose and achievements. Green communities function as 'learning institutions'. This is especially evident in the case of Citizens House, which benefitted from a national and regional framework of broad-based community organising, following a theory of change that Lewisham Citizens implemented as a learning institution. The groups and projects identified in Box 1.1 can all be viewed as social learning systems because people are working and learning together without any formal educational expectations.

Social learning is characterised by open communication and informal, unrestricted participation, as a function of observation and imitation and learning from each other in social situations that are endowed with mutually agreed purpose, e.g., picking up litter, supervising children at play, and visiting another group or project. Mark Reed and his colleagues argue that for social learning, a process must demonstrate that a change in understanding has taken place in the individuals involved, that this change becomes embedded in wider communities of practice, and that it occurs through interactions cultivated by this community of practice (Reed et al. 2010: 1). Social learning is not a linear or one-way process. Individual and collective intentions vary in resolve such that they ebb and flow with changing experience of conscious listening and dialogue. This is what makes them dynamic and also aligns with the multi-level approach outlined above. Social learning illustrates the potential to build an alliance for the diffusion of innovation as a theory of change, to challenge the dominant regime in order to steer placemaking in a just direction.

Concluding arguments: Effective collective action

This chapter draws attention to new and retrofitted social architectures of hyper-local micro-activism to align communities that are organising to deliver progressive placemaking according to existing research on intentional communities. These are important stories of people, place, and power that are largely overlooked in official accounts of placemaking. Missing from the preoccupation with investments in city-wide social infrastructure and the attitudes, behaviour, and choices of individual consumers are stories of grassroots groups that are sufficiently well-organised to amplify their collective power in pursuit of a common goal.

Questions arise from the discussion above about the way that grassroots groups build alliances with other stakeholders in order to access dominant regime support. This discussion highlights the craft skills and social learning that are necessary for niche innovations to demonstrate successful *practivism* (Hoff et al. 2020). There is much we can learn from networks of niche innovation. However, we must take care not to overlook diverse cultural settings, whether in the UK, Northern Europe, or indeed, in informal urbanism in the Global South. The social scale relations of sharing are complex, fragile, and difficult to contrive. In many cases, the appropriate role for local government is to get out of the way of green communities to ensure their successful organic growth (Moore et al. 2015).

This largely theoretical chapter began with a critique of existing literature and case studies on placemaking where the language of public participation assumes a measure of democracy and fairness without theorising how people gather and organise in practice. It describes a dialogue of personal transformation and collective social action (Braidotti 2020). Key to achieving a just transformation in locally determined solutions to global crises are harmonious human-environmental connections and a unity of purpose.

Arguably, alongside a multi-level approach, a more generative vocabulary that is more socially conscious is required to account for the ambiguities and architectures of a negotiated 'we' of interpersonal connection and shared purpose. All too often, conventional typologies of community-as-neighbourhood, voluntary organisation, network, hub, or locale are reduced to territorial or institutional distinctions by size and constitutional status. We need a nuanced analysis, vocabulary, and engaged methods that better understand and ultimately share the 'secret' constituents of human-environmental connectedness if we are to make just transitions scalable. We need to better understand what forms of support help amplify innovation diffusion as a recommendation for planning and policy, to replace or qualify the current presumption that determines success in terms of building and funding meeting spaces.

This is not to say that we do not need spaces of congregation and common interest. Libraries not only provide inclusive access to spaces of togetherness, they also serve an essential role in affordable access to social learning (books, toys, digital technologies, and skills). However, unless attention is paid to the 'soft' social architectures of relational power, these will remain sites of impersonal transaction, limited encounters, and consumer citizenship. It is not possible to engineer relationships of trust and social change through the design and layout of the built environment alone. It is telling from the five examples compared in Box 1.1 that it is the participatory *practivism* of the Dwellbeing cooperative that cultivates a sense of community and belonging in Shieldfield, not the physical features of a concrete landscape that can appear unloved and hostile to the first-time visitor.

The multi-level approach suggested here, which constitutes niche, regime and landscape, is one way to account for the way groups and projects navigate complex interdependencies in pursuit of sustainable degrowth goals in the here and now. Contrary to the pejorative view of the IC as an island of escape, 'people-powered' placemaking is typically pragmatic rather than idealistic. It pursues a vision with purpose where the realities of being part of a larger social learning system (a community of communities) (Illich 1973) help reduce isolation and prevent participants' burnout. Arguably, groups must attend to their own collective well-being if they are to meaningfully connect local to global structures and institutions, here-and-now and over the long-term, which includes challenging structural barriers to carbon reduction, affordability, diversity, and inclusion. In this, we observe a paradox. The intimate scale of belonging and trust that enable intentional communities of practice to thrive can also unintentionally inhibit efforts to challenge structural inequalities, notably racism, and an effective theory of change.

From a critical scrutiny of economic growth measured by Gross Domestic Product (GDP), we learn that social economies of sharing (inhibited in a dominant regime of urban competitiveness) thrive on the conviviality of degrowth activism. The 'tools' of conviviality are difficult to access in virtual communities or in dispersed communities where real-time in-person interactions are infrequent. Nevertheless, togetherness and trust can be about 'being present' in ways that are not predicated on enduring co-location. This is evident in the practical examples above and in new forms of broad-based community organising today that deepen the democracy and power of civil society (Citizens UK 2024). It is possible to observe and cultivate tools of conviviality and architectures of cooperation on multiple levels; this task is not limited to the communal activism of intentional eco-communities. Significantly, it is useful to conceptualise the conditions under which the glue of interpersonal trust and purpose thrive as a 'community of communities' (Illich 1973). By situating activists and hyper-local activism within multi-level and broad-based alliances of grassroots placemaking, we can reclaim diverse local expertise, social learning, and ecological vernaculars and transform potential power into actual power to act on the tangible issues of system change (Hinchliffe & Whatmore 2006: 134).

References

Bergman, N., Whitmarsh, L., Köhler, J., Haxeltine, A., & Schilperoord, M. (2007). Assessing transitions to sustainable housing and communities in the UK. In M. Horner, C. Hardcastle, A. Price, & J. Bebbington (Eds.), *International conference on whole life urban sustainability and its assessment* (pp. 1–31). Glasgow.

Bhargava, D., & Luce, S. (2023). *Practical radicals: Seven strategies to change the world*. The New Press.

Blank, J. (2001). Common meals in cohousing communities: An experienced cohouser reports from common house kitchens and dining rooms across North America. *Cohousing Journal*, Winter 2001, http://joaniblank.com/cohousing/CommonMeals/index.html

Boyer, R. H. W. (2015). Grassroots innovation for urban sustainability: Comparing the diffusion pathways of three ecovillage projects. *Environment and Planning* A, 47(2), 320–337. https://doi.org/10.1068/a140250p

Braidotti, R. (2020). 'We' are in this together, but we are not one and the same. *Journal of Bioethical Inquiry*, 17(4), 465–469.

Bunce, S. (2020). Engagement and activism in community land ownership: The emergence of community land trusts in London and Toronto. In S. Bunce, N. Livingstone,

L. March, S. Moore, & Walks (Eds.), *Critical dialogues of urban governance, development, and activism* (pp. 274–288). University College London.
Chambers, E. T. (2017). *The power for relational action*. ACTA Publishing.
Chatterton, P. (2023). *How to save the city: A guide for emergency action*. Agenda Publishing.
Christian, D. L. (2003). *Creating a life together: Practical tools to grow ecovillages and intentional communities*. New Society Publishers.
Citizens UK (2024). Theory of change model: Access organization values. www.citizensuk.org/about-us/our-values/
Citizens UK (2025). Citizens UK Manifesto. www.citizensuk.org/campaigns/citizens-manifesto/
Civic Square (Birmingham) (2023). Demonstrating neighbourhood-scale civic infrastructure for social, ecological, economic, and climate transition. https://civicsquare.cc/ (other publications available online).
Constantino, S. M., Sparkman, G., Kraft-Todd, G. T., Bicchieri, C., Centola, D., Shell-Duncan, B., Vogt, S., & Weber, E. U. (2022). Scaling up change: A critical review and practical guide to harnessing social norms for climate action. *Psychological Science in the Public Interest*, 23(2), 50–97. https://doi.org/10.1177/15291006221105279
D'Alisa, G., Demaria, F., & Kallis, G. (Eds.) (2015). *Degrowth: A vocabulary for a new era*. Routledge.
Daly, M. (2018). *The most powerful form of activism is just the way you live: Grassroots intentional communities and the sustainability of everyday practice*. [Dissertation, University of Technology Sydney]. University of Technology Sydney.
Donaldson, G. H., & João, E. M. (2020). Using green infrastructure to add value and assist place-making in public realm developments. *Impact Assessment and Project Appraisal*, 38(6), 464–478.
Forrest, R., Murie, A., & Williams, P. (2021). *Home ownership: Differentiation and fragmentation*. Routledge.
Geels, F. W., & Schot, J. (2007). Typology of sociotechnical transition pathways. *Research Policy*, 36(3), 399–417.
Gehl, J., & Svarre, B. (2013). *How to study public life*. Island Press.
Hayden, A. (1999). *Sharing the work, sparing the planet: Work, time, consumption and ecology*. Zed Books.
Heywood, A. (2016). *Local housing, community living: Prospects for scaling up and scaling out community-led housing*. The Smith Institute. www.bridport-tc.gov.uk/wp-content/uploads/2019/04/local-housing-community-living.pdf
Hinchliffe, S., & Whatmore, S. J. (2006). Living cities: Towards a politics of conviviality. *Science as Culture*, 15, 123–138.
Hoff, J., Gausset, Q., & Westergaard, S. L. (2020). Introduction. In J. Hoff, Q. Gausset, & S.L. Westergaard (Eds.), *The role of non- state actors in the green transition* (pp. 1–14). Routledge.
Illich, I. (1973). *Tools for conviviality* (2009 ed.). Marion Boyers.
Jarvis, H., Pratt, A. C., & Wu, P. C. C. (2001). *The secret life of cities: Social reproduction of everyday life*. Routledge.
Jarvis, H. (2011). Saving space, sharing time: Integrated infrastructures of daily life in cohousing. *Environment and Planning A*, 43(3), 509–762.

Jarvis, H. (2015). Towards a deeper understanding of the social architecture of cohousing: Evidence from the UK, USA and Australia. *Journal of Urban Research and Practice*, 8(1), 93–105.

Jarvis, H. (2018). Envisioning liveability and do-it-together urban development. In A.B. Jonas, B. Miller, & K. Ward (Eds.), *Handbook of spaces of urban politics* (pp. 336–349). Routledge.

Jarvis, H. (2019). Sharing, togetherness and intentional degrowth. *Progress in Human Geography*, 43(2), 256–275.

Jarvis, H. (2023). Community organising in higher education: Activist community-engaged learning in geography. *Journal of Geography in Higher Education*, 48(3), 368–388.

Jarvis, H. (2025). How eco-communities grow through social learning, social permaculture, and group transformation. In J. Pickerill (Ed.), *Eco-communities: Surviving well together* (pp. 227–239). Bloomsbury.

Jarvis, H., & Bonnett, A. (2013). Progressive nostalgia in novel living arrangements: a counterpoint to neo-traditional new urbanism? *Urban Studies*, *50*(11), 2349–2370.

Lang, R., Chatterton, P., & Mullins, D. (2020). Grassroots innovations in community-led housing in England: The role and evolution of intermediaries. *International Journal of Urban Sustainable Development*, 12(1), 52–72.

Latham, A., & Layton, J. (2022). Social infrastructure: Why it matters and how urban geographers might study it. *Urban Geography*, 43(5), 659–668.

Litfin, K. T. (2014). *Ecovillages: Lessons for sustainable community*. Polity.

Ludwig, M. (2017). *Together resilient: Building community in the age of climate disruption*. Routledge.

Moore, M. L., Riddell, D., & Vocisano, D. (2015). Scaling out, scaling up, scaling deep: Strategies of non-profits in advancing systemic social innovation. *Journal of Corporate Citizenship*, 58, 67–84.

Morgan, D. H. (2020). Family practices in time and space. *Gender, Place & Culture*, 27(5), 733–743.

Nelson, A. (2018). *Small in necessary: Shared living on a shared planet*. Pluto.

Nichols, K. (2024). Starting a new cohousing group. *Communities*, 203, 28–31.

Ochre, G. (2013). *Getting our act together: How to harness the power of groups*. Groundworks Press.

Power to Change (2025). *York supplies: Continuing the legacy of a 'kind place' to shop*. Cooperatives UK case study. www.uk.coop/case-studies/york-supplies-continuing-legacy-kind-place-shop-cs

Ray, S. J. (2020). *A field guide to climate anxiety: How to keep your cool on a warming planet*. University of California Press.

Reed, M. S., Evely, A. C., Cundill, G., Fazey, I., Glass, J., Laing, A., Newig, J., Parrish, B., Prell, C., Raymond, C., & Stringer, L. C. (2010). What is social learning? *Ecology and Society*, 15(4). www.jstor.org/stable/26268235

Richards, G. (2021). Creative placemaking strategies in smaller communities. In K. Scherf (Ed.), *Creative tourism in smaller communities: Place, culture, and local representation* (pp. 283–298). University of Calgary Press.

Sanford, W. (2013). Being the change: Gandhi, intentional community and the process of social change. *Social Sciences Directory*, 2(3), 106–113.

Sanne, C. (2002). Willing consumers—or locked-in? Policies for a sustainable consumption. *Ecological Economics*, 42(1–2), 273–287.

Sara, R., Jones, M., & Rice, L. (2020). Austerity urbanism: Connecting strategies and tactics for participatory placemaking. *CoDesign*, 17(4), 493–509. https://doi.org/10.1080/15710882.2020.1761985

Schalk, M., Gunnarsson-Östling, U., & Bradley, K. (2017). Feminist futures and 'other worlds': Ecologies of critical spatial practice. In S. MacGregor (Ed.), *Routledge handbook of gender and environment* (pp. 447–463). Routledge.

Schiller-Merkens, S. (2024). Prefiguring an alternative economy: Understanding prefigurative organizing and its struggles. *Organization*, 31(3), 458–476.

Sennett, R. (2012). *Together: The rituals, pleasures and politics of co-operation.* Allen Lane.

Sennett, R. (2012a). The architecture of co-operation. In *Architecture, Planning and Landscape Annual Lecture, 13 December 2012*, Newcastle University.

Shove, E. (2010). Beyond the ABC: Climate change policy and theories of social change. *Environment and Planning A*, 42(6), 1273–1285.

Tanner, W., O'Shaughnessy, F., Krasniqi, F., & Blagden, J. (2024). The state of our social fabric: Measuring the changing nature of community over time and geography. Report. Onwards. www.The-State-of-our-Social-Fabric.pdf

Wenger, E. (2000). Communities of practice and social learning systems. *Organization*, 7(2), 225–246.

Wright, E. O. (2013). Real utopias. *Politics & Society*, 41(2), 167–169.

2

EROS UNREDEEMED

Post-patriarchal soft power and inclusivity in the Tamera and Monte Verità communities

Ana Margarida Esteves

Introduction

Inclusivity is paramount for the scalability of green communities as relevant models for sustainability transition. However, those promoted by vanguards prioritising behavioural change over the communing of material resources tend to become niches reserved for select and socially homogeneous groups. This creates dynamics that often reproduce exclusionary patterns and hinder long-term scalability and structural inclusivity, which compromises the survivability of these projects. Intentional communities like Tamera, Portugal or Damanhur, Italy, for instance, often rely on unpaid trial periods, fluency in specific cultural idioms, or the capacity to live without waged income for extended periods—factors that filter out those lacking class privilege or transnational mobility.

This chapter uses a comparative case study analysis to explore how green intentional communities that prioritise cultural innovation deploy their counter-cultural status as a form of post-patriarchal soft power—a symbolic resource grounded in aesthetic dissent from patriarchal, industrial modernity.

The concept of post-patriarchal soft power refers to the use of emotional, ethical, and symbolic capital rather than coercion or institutional authority to cultivate legitimacy and mobilise support. Drawing from Joseph Nye's (2017) notion of soft power as 'the ability to obtain preferred outcomes by attraction rather than coercion or payment' and extending Pierre Bourdieu's (1986) idea of symbolic capital, post-patriarchal soft power operates through the projection of alternative values such as care, sensuality, and ecological harmony as desirable and morally superior. Unlike corporate or state soft power, which often relies on mass media and branding, post-patriarchal

DOI: 10.4324/9781003769064-3

soft power is enacted through embodied practices, lifestyle performances, and pedagogical experiences within the community. It also resonates with Max Weber's (1978) concept of charismatic authority, in which legitimacy is conferred upon individuals or collectives that are perceived to possess extraordinary moral or spiritual qualities. In this sense, Tamera and Monte Verità position themselves not only as experimental societies but as ethical vanguards in a global system perceived as alienating, destructive, and patriarchal.

In these contexts, soft power is often expressed through values such as emotional transparency, ecological harmony, and liberated sexuality. These ideals are projected outward to attract donors, political allies, and aspiring members who are culturally aligned with the community's vision. Drawing on Bourdieu's (1986) concept of social capital, the chapter explores how such attraction generates durable networks of support among elite publics. However, this symbolic capital also tends to reproduce barriers to access, especially for individuals without the economic means to cover the expenses of unpaid provisional membership periods, or the cultural fluency to navigate unfamiliar norms and languages. As a result, communities like Tamera may unconsciously replicate social hierarchies even while explicitly rejecting them.

Through a comparative analysis of Monte Verità (1900–1926) and Tamera (1994–present), the chapter explores how post-patriarchal soft power is deployed to engage public and institutional actors. It examines how countercultural experiments in social regulation, particularly around gender, sexuality, and nature, generate symbolic appeal yet often fail to overcome classed, cultural, and institutional barriers to inclusivity and scalability. The analysis is guided by two key questions: how is soft power produced and circulated within green intentional communities? And what are its unintended consequences for access, diversity, and long-term viability?

The chapter ultimately considers how alternative governance models might better integrate structural inclusivity without losing cultural distinctiveness. The findings highlight the need for green communities and the public sector to partner in developing mechanisms for economic and cultural inclusivity, including regulation for commons-based governance and intersectional solidarity with broader social movements.

Cultural and structural determinants of green intentional communities

This chapter builds on the concept of soft power that was originally developed by Nye (2017) to describe the ability to influence others through attraction and legitimacy rather than coercion or monetary incentive. In the context of green intentional communities, soft power functions as a form of symbolic capital (Bourdieu 1986): it legitimises and disseminates ethical lifestyles organised around ecological sustainability, emotional transparency, and gender/sexual liberation.

To move beyond Nye's geopolitical framing, this chapter draws on complementary critical approaches. J. K. Gibson-Graham's (2013) work on community economies emphasises the need to root economic alternatives in everyday relational practices. Silvia Federici's (2018) feminist critique of the commons highlights how idealised models can become detached from the material realities of care and social reproduction. Nancy Fraser's (1997) dual lens of redistribution and recognition reveals the tension between cultural experimentation and structural exclusion. Together, these perspectives help illuminate how green intentional communities generate appeal while simultaneously reproducing selective patterns of belonging.

This tension is particularly visible in the way communities project post-patriarchal values as transformative models. While symbolically potent, these models often encode classed, racialised, and geographic boundaries. Cultural innovation can become an exclusionary force when it is shaped by aesthetic and communicative norms that are accessible only to a narrow demographic.

The literature on alternative modernities further clarifies this dynamic. Green intentional communities typically draw on two overlapping critiques of industrial society. One, rooted in anti-industrial romanticism, promotes lifestyle reform through communal living, vegetarianism, and nature-based healing (Sirost 2016; Mongini, Gatti, & Risaliti 2023). The other, informed by political economy, seeks structural transformation via commons-based governance and participatory decision-making (Forno & Wahlen 2022). Forno and Wahlen's research on urban environmental activism shows how horizontal governance structures and collaborative consumption practices anchor more inclusive and materially grounded alternatives.

Which critique is prioritised by a given community often depends on its socioeconomic and geopolitical positionality. Indigenous-led or postcolonial communities in Latin America and southern Africa, for example, typically centre communal landholding and food sovereignty as strategies of material survival (Escobar 2018). By contrast, communities initiated by globally mobile, materially secure actors such as Tamera or Damanhur tend to emphasise symbolic protest and cultural experimentation (Esteves 2017; Pinto & Vilaça 2023; Mongini, Gatti, & Risaliti 2023).

Despite these differences in orientation, green intentional communities face shared structural constraints. One major limitation is the lack of public regulation adapted to commons-oriented governance (Wolsink 2020). For instance, Findhorn ecovillage in Scotland has struggled to reconcile cooperative land management with municipal planning regimes.

Market pressures to scale and replicate also shape community strategies: Zentrum für experimentelle Gesellschaftsgestaltung (ZEGG), an intentional community in Germany, for example, increasingly relies on commodified educational retreats to ensure economic viability (Seyfang, Smith, & Longhurst 2010). Moreover, post-extractivist approaches to self-sufficiency often limit a community's capacity to generate capital surpluses or achieve economies of scale (Altvater 2005; Chatterton & Pickerill 2010). This is particularly true for communities that rely on revenue from retreats, permaculture courses, or wellness tourism, whose pricing structures cater to affluent participants (Pinto & Vilaça 2023). In such cases, ecological resilience is dependent not only on material design but also on shifting donor landscapes and policy climates (Walker & Devine-Wright 2008).

Research has shown that even as these communities advance cultural innovation, they often reproduce exclusionary dynamics like those found in mainstream society (Turner 2006; Pickerill & Maxey 2009). Communities like Tamera or Findhorn, for example, typically require that members align with specific ethical values, the capacity to work without pay during initiation periods, or the ability to afford costly workshops—criteria that function as implicit filters privileging urban, white, Northern European participants (Meijering, Huigen, & van Hoven 2007).

Importantly, these exclusions are rarely the result of deliberate policy. Rather, they emerge through subtle, embodied practices: dominant linguistic norms (e.g., English or German), aesthetic preferences, and emotional communication styles rooted in Northern European therapeutic cultures (Tollefson 1991; Melucci 1996; Kunze 2012). These features help build internal cohesion but can also make others feel culturally out of place. In Bourdieu's terms, they reflect a particular *habitus*—a system of internalised dispositions that shapes perception and behaviour (Bourdieu 1984). Following Michel Foucault (1977), we can also understand these practices as forms of diffuse power: they are not imposed from above but are reproduced through everyday interaction. These mechanisms of cultural gatekeeping complicate otherwise sincere attempts to promote diversity and inclusivity, which reveals the contradiction between universalist ideals and exclusive material and symbolic infrastructures.

Case studies

The analysis presented in this chapter focuses on two green intentional communities that share overlapping cultural aims but are situated in very different historical and geographical contexts.

Tamera is a living community founded in 1994 in the Alentejo region of southern Portugal. It identifies as a 'Peace Research & Education Center' and currently hosts around 170 permanent residents and a steady flow of visitors, students, and international guests. Its founders, Dieter Duhm and Sabine Lichtenfels, emerged from the German academic and countercultural milieu.

Today, Tamera functions as a site of ecological experimentation, spiritual inquiry, and post-capitalist innovation.

The title of this chapter, 'Eros Unredeemed', is inspired by Duhm's seminal work, *Eros Unredeemed: The World Power of Sexuality* (Duhm 2010), in which he outlines for Tamera a vision for post-patriarchal social transformation grounded in the reintegration of love and sexuality into the ethical and ecological fabric of community life.

Monte Verità, by contrast, was founded in 1900 on Monescia Hill overlooking the town of Ascona in the Swiss canton of Ticino. The vision of founders, Austro-Hungarian pianist Ida Hoffmann and Henry Oedenkoven, the son of a Belgian industrialist, was shaped by the Central European countercultural currents of their time. Though it thrived briefly as a haven for artists, anarchists, and Lebensreform practitioners, the community disbanded in 1926 when the property was sold to a private investor. What remains is a rich archival legacy preserved in the Monte Verità Foundation and in exhibitions curated by the late Swiss artist Harald Szeemann.

Both communities were founded in peripheral regions that, at the time, lay at the cultural and infrastructural margins of Europe. Monte Verità was located in Ticino, an Italian-speaking canton bordering Lombardy, considered at the turn of the 20th century to be culturally part of Southern Europe. The region lacked a railway connection to other parts of Switzerland or Central Europe. According to contemporary accounts, travelling from Germany or Austria-Hungary to the site took at least three days, with the final stretch through the southern Swiss Alps completed on foot (Mongini, Gatti, & Risaliti 2023).

Similarly, Tamera was established in the municipality of Odemira, one of the largest but also most socioeconomically disadvantaged regions in Portugal. At the time of this study's fieldwork in 2015, it was among the poorest and most sparsely populated areas in the European Union, shaped by long-term rural depopulation and a process of environmental and economic desertification throughout the 20th century (Esteves 2017: 980).

In both cases, remoteness and marginality created space for cultural experimentation by socially mobile outsiders, while also shaping the practical constraints and social dynamics they would later face.

The ideological roots of both communities can be traced to a wider philosophical and spiritual genealogy that runs through German Idealism, *Romantische Naturphilosophie*, and esoteric cosmology. As Sirost (2016: 29) observes, the utopian spirit animating Monte Verità was nurtured by a lineage that stretched from 18th-century theosophy to the Dionysian and bohemian circles of 19th-century Europe. Thinkers like Franz von Baader, Friedrich Hölderlin, Friedrich Schelling, and Georg W. F. Hegel contributed to a cosmology that merged inner energetics, natural mysticism, and a reinvestment in sensuality and pagan ritual.

These ideas shaped artistic subcultures across Europe—from Mallarmé's Mardistes to Munich's Schwabing Cosmic Circle—and ultimately laid the imaginative groundwork for communities like Monte Verità and, indirectly, Tamera. It was within these alternative cosmologies that many countercultural movements sought to reforge the relationship between the individual, the body, and the natural world in ways that challenged both Enlightenment rationalism and institutional religion.

Both communities were made possible by the freedoms afforded to their founders: geographical and cultural distance from urban power centres and the financial autonomy that came with being economically self-reliant outsiders (Esteves 2017; Mongini, Gatti, & Risaliti 2023). These conditions enabled the creation of intentional communities in peripheral regions that were relatively shielded from institutional scrutiny.

However, this relative autonomy also contributed to a cultural and demographic disconnect from their surroundings. The socioeconomic homogeneity of residents and long-term visitors reinforced this divide, attracting individuals from similar class and cultural backgrounds while limiting wider integration. These peripheral settings functioned as what one might call cultural no man's lands—zones of autonomy that attracted those seeking to live outside the reach of dominant institutions (Turner 2006; Scott 2009).

Despite their differing lifespans, both communities drew inspiration from a shared ideological lineage rooted in German Romanticism and Idealism. This legacy was transmitted through the *fin-de-siècle* Lebensreform movement, which advocated natural living, body culture, and holistic reform in response to industrial modernity, and re-emerged in the late 20th-century back-to-the-land and eco-spiritual movements. These traditions articulate a persistent critique of industrial modernity, emphasising the suppression of emotion, embodied experience, and ecological interdependence (Sirost 2016).

In Tamera, this legacy is embodied in the concept of Eros as a foundational principle of social organisation. Eros, in Duhm's framing, extends beyond sexuality to encompass the energetic forces of connection, reciprocity, and vitality. His book *Eros Unredeemed* outlines a model for a post-patriarchal society in which cognitive, emotional, and somatic life has been long fractured by rationalism and control are reintegrated into a cooperative and ecologically attuned social order.

This shared framework informed both communities' emphasis on behavioural transformation, including:

- Rejection of patriarchal and hierarchical social norms;
- Emphasis on emotional transparency and collective healing;
- Daily practices aligned with ecological cycles and spiritual insight; and

- Commitment to prefigurative politics, i.e., living as if the desired future were already present.

Yet the communities experienced markedly different trajectories. Monte Verità was established in the very early 20th century, during a time of escalating geopolitical conflict among European empires. While its location in neutral Switzerland offered refuge for dissidents and artists, Swiss neutrality and the country's status as a banking haven could not shield the community from the wider economic and political shocks of World War I and the interwar period (Mongini, Gatti, & Risaliti 2023). It remained largely insular and made little effort to engage with the local population. After two decades, it succumbed to internal fragmentation and financial precarity.

Tamera, by contrast, emerged in the 1990s, a period of relative peace and institutional stabilisation, and was shaped by Portugal's democratic transition and integration into the European Union. These structural conditions allowed Tamera to develop long-term strategies for social legitimacy, transnational alliances, and forms of engagement with surrounding villages and public authorities. Over time, the community cultivated relationships with former Portuguese emigrants who acted as cultural mediators and helped bridge linguistic and ideological divides. In 2024, Tamera celebrated its 30th anniversary as an active and evolving community (Esteves 2017).

Yet historical context alone cannot fully explain these divergent outcomes. This chapter argues that the ways in which each community deployed cultural innovation, particularly through what I term post-patriarchal soft power, also played a decisive role. Symbolic appeals to love, nature, and freedom served as strategies of attraction that helped secure material support, institutional legitimacy, and social recognition. As the following sections show, these strategies simultaneously expanded the reach of each community and reinforced new forms of exclusion and selectivity.

Methodology

This chapter uses a comparative-historical ethnographic approach to examine the cases of Monte Verità and Tamera. These communities, founded nearly a century apart, are studied not to draw empirical parallels or assess outcomes, but to trace ideological continuities and strategic patterns in how green intentional communities mobilise soft power, attract symbolic and material capital, and manage the dynamics of inclusion and exclusion over time.

Comparing communities separated by over a century raises valid concerns around historical discontinuity and the risk of anachronism. However, this chapter uses a genealogical approach, inspired by Foucault (1977), not to track linear causality but to surface the recurrence of cultural logics such as post-patriarchal critique and nature-based ethics within structurally distinct moments. These cases are connected not by institutional lineage but by ideological resonance. The comparative method used here follows George Marcus's (1995) model of diachronic, multi-sited ethnography: understanding how parallel experiments in social innovation are shaped by their historical settings while drawing from overlapping repertoires of dissent. This also allows us to examine how countercultural actors have continuously tried to negotiate visibility, legitimacy, and internal cohesion without dissolving into mainstream norms.

This approach aligns with Foucault's (1977) genealogical method, which seeks to uncover the layered and often discontinuous histories of concepts and practices. It is also informed by Reinhart Koselleck's (2004) work in conceptual history, which foregrounds how terms like 'community', 'freedom', or 'nature' shift in meaning across historical contexts and political regimes. These shifts help explain how both Monte Verità and Tamera, despite structural differences, utilise cultural innovation to attract like-minded publics while navigating constraints shaped by their respective eras.

The rationale for the diachronic comparison emerged during my initial research project on contemporary eco-communities and networks of transition initiatives in Southern Europe. A key insight of this project, which was developed in the article 'Commoning at the Borderland' (Esteves 2017), was that inclusion and exclusion within these communities were often shaped by cultural and classed expectations that remained invisible to their own members. This raised the question of whether such patterns were historically contingent or embedded in longer-standing cultural logics. Returning to earlier projects like Monte Verità helped clarify how strategies of attraction—particularly those aimed at securing financial viability through cultural capital—have been central to community survival for over a century. However, the historical contexts matter: while Monte Verità appealed to a global elite in search of refuge and aesthetic innovation, Tamera's growth has been shaped by processes of globalisation, including rural exodus due to shifts in transnational agricultural policy and supply chains, the global circulation of countercultural ideas through information technologies, and the increasing popularity of southern Portugal as a destination for tourism and expatriation by the middle and upper classes of industrialised countries.

My fieldwork in Tamera was carried out between April and November 2015 and consisted of daily participant observation, informal conversations, and 15 semi-structured interviews with community members. I also took notes of conversations with neighbours, regional policymakers, and civil society actors. Preliminary analysis suggested that Tamera has an ideological and strategic continuity with Monte Verità, which Tamera's founders have explicitly referenced as a cultural and philosophical ancestor. This is especially evident in Tamera co-founder Sabine Lichtenfels's memoir, *Traumsteine. Reise in das zeitalter der sinnlichen erfüllung* (*Dreamstones. Journey into the age of sensual fulfilment*), where she situates Tamera within a lineage of sensual, ecological, and spiritual renewal connected to Monte Verità (Lichtenfels 2000).

Beginning in September 2024, I started a follow-up research project on the historical roots of the ecovillage movement. Data collection began with archival work at the Monte Verità Foundation. This included research in the multimedia archives curated by Harald Szeemann that are located at the Museum Casa Anatta, the Elsarion Pavilion, the Maggia municipal archive, and the library of the Bauhaus Hotel in Ascona. The material consulted—primarily in German, Italian, French, and English—includes diaries, correspondence, newspaper articles, and administrative documents from community members, regional authorities, and visitors. The documents compiled for Szeemann's Mammelle della Verità Exhibition (1978) and his role in Monte Verità SA's cultural commission (1984–1990) provided rich context on how the community framed its public image.

Data analysis followed a process of triangulation across these diverse sources to integrate integrating ethnographic observation with archival content. A deductive coding scheme was then applied to identify recurring themes across the two field sites, including aesthetic regimes, narratives of transformation, and implicit mechanisms of social selection. This method draws on Marcus's (1995) framework of multi-sited and diachronic ethnography, which enables comparative insight into how cultural formations persist or evolve across distinct temporal and spatial configurations.

Transitioning beyond patriarchy: Eros and social regulation

During a conversation with a fellow scholar-activist, I asked what motivates people from affluent societies to seek out life in intentional communities. His reply captures a key dynamic behind the emergence of both Tamera and Monte Verità:

> People in intentional communities, especially in Germany and other countries in central and northern Europe (...) experienced the best that capitalist society can offer and are not happy about it. They are unsatisfied and long for something else. (...) Their cultural capital is so high that they see the limitations, the pitfalls of what capitalism has to offer to those who benefit the most from it. That's why they come together. (...) Because of their cultural capital, they wouldn't do it otherwise.[1]
>
> *(Private conversation during fieldwork, October 2024)*

This insight speaks to the central motivation of both communities: the desire to create a container for alternative ways of living, i.e., experiments in daily life that challenge the dominant values of capitalist modernity. In this sense, both Tamera and Monte Verità are and were cultural laboratories for what their founders envisioned as an alternative modernity, one not based on accumulation, rationalism, and control, but on interdependence, emotional intelligence, and ecological attunement.

At the core of this project lies a reconceptualisation of the human relationship's connection to the other-than-human world—not as a resource, but as a network of sentient and responsive life. This perspective resembles worldviews found in indigenous cosmologies, animist traditions, and polytheistic philosophies such as Hinduism, which treat nature as alive and consciousness as distributed across bodies, beings, and environments. In both communities, this reconnection is and was pursued through a rebalancing of rational cognition with the somatic and emotional dimensions of perception and meaning making (Esteves 2017).

This transformation is organised around the principle of Eros, which both communities elevate as a foundation for social regulation. Eros here refers not only to sexuality, but to the broader spectrum of cognitive, emotional, and somatic experiences of pleasure—the kind of embodied reciprocity that emerges from meaningful connection between beings (Sirost 2016; Russo 2024). In both Tamera and Monte Verità, Eros is treated as a social force with political implications. As the writings of contemporaries such as Otto Gross (1913), Erich Mühsam (1999, 2002, 2002a), and Gustav Landauer (2009) attest, many early 20th-century utopians saw capitalism and industrial modernity as ultimate expressions of patriarchal norms. For them, organising public morality and social regulation around Eros was inseparable from the abolition not only of gender hierarchies, but, above all, of class society.

This framework informs the everyday practices of both communities, based on emotional transparency, therapeutic engagement, and ecological embeddedness. Practices include art therapy, psychoanalytical techniques, herbal medicine, homeopathy, physical movement, and a vegetarian diet. These are accompanied by a wellness regime designed to realign human bodies and emotions with natural cycles. Underpinning this is an ethos of voluntary simplicity and reciprocity: frugality in consumption, localisation of food and goods, and emphasis on cooperation over competition.

In this context, both communities prioritise behavioural change over formal structural transformation. Although inspired by the idea of commons, neither Monte Verità nor Tamera practiced full economic collectivism. The land and infrastructure of Monte Verità were privately owned by the founding couple, Ida Hoffmann and Henry Oedenkoven (Mongini, Gatti, & Risaliti 2023). In Tamera, property is held by ILOS, Centro de Pesquisa para a Paz, Lda., a private company that manages donations and income from educational programmes. Its shareholders are non-profits: The Grace Foundation and the Associação para um Mundo Humanitário (Association of a Humanitarian World) (AMH), but communal ownership and income-sharing are absent.

In this context, 'commoning' refers more to cooperative ethos and shared intention than to legal-economic arrangements. It involves building trust-based relations of interdependence and co-responsibility that are often enacted through shared tasks, rituals, and care work, rather than the collectivisation of material assets.

Since its founding, Tamera has placed particular emphasis on the liberation of human sexuality from patriarchal norms. Practices such as polyamory, transparency in intimate relationships, and 'love schools' form a central part of its internal culture. While Monte Verità was also concerned with sexual liberation, it treated sexuality as one among many facets of a broader cultural project. In both cases, however, Eros functions as a tool for unlearning the emotional and relational damage inflicted by industrial modernity and as a path toward a society rooted in care, connection, and ecological reciprocity.

Cultural innovation, economic selectivity, and local tensions

> I am the black sheep of a wealthy family. It is natural that I will feel resonance with other black sheep of wealthy families, and that we will want them to come here.
>
> *(Interview with Tamera community member, September 2015)*

> They do not live in Portugal. They live in a world apart. They want to develop the project of two people from scratch. (...) They touch on a lot of questions that are taboo over here.
>
> *(Local resident, fieldwork notes, October 2015)*

These comments, one by a long-term member of Tamera, the other by a Portuguese neighbour, reveal the cultural and economic divide between the community and its surrounding population. They also highlight the selective nature of participation in green intentional communities that prioritise cultural innovation over the development of a self-sustaining, locally embedded economy. These selective pressures shape who feels 'at home' in such settings, and who is excluded, both materially and symbolically.

Tamera's cultural model, with its emphasis on liberated sexuality and spiritual ecological design, not only challenges mainstream values but often appears illegible to local observers. In James Scott's (2009) terms, it lacks 'legibility' in the eyes of both the state and traditional communities. While the community's internal logic, based on healing trauma, emotional literacy, and planetary care, is coherent to its members, it registers externally as eccentric, opaque, or even threatening. This creates a double bind: the very cultural distinctiveness that sustains internal cohesion and symbolic appeal may obstruct social integration. The resulting tension exemplifies the dilemma of cultural innovation within countercultural experiments: the pursuit of transformative ideals often risks reinforcing isolation unless strategic cultural translation efforts are made.

Both Tamera and Monte Verità attracted a similar demographic: predominantly upper- and upper-middle-class Central Europeans seeking expatriation not for integration, but to build alternative forms of life from the ground up. These individuals are not outcasts in the traditional sense, but pioneers—culturally privileged actors with the economic and symbolic capital to critique the very systems from which they benefitted.

Joining such communities implies a level of independence from the labour market. It presumes access to inherited wealth, property rental income, or social safety nets capable of absorbing the economic risks of countercultural experimentation. As a result, community membership tends to favour those whose backgrounds resemble that of the founding cohort, which fosters strong internal affinity but also reinforces class boundaries.

Monte Verità's early economic model exemplified this pattern. The founders, Ida Hoffmann and Henry Oedenkoven, financed the infrastructure and invited like-minded participants to contribute to utility costs and upkeep. Yet even modest monetary requirements led to ideological tensions. Founding member Gusto Gräser left the community early to seek a more radical anti-economic lifestyle.

Over time, Monte Verità shifted toward monetising natural medicine and cultural programming, transforming itself into a sanatorium offering holistic therapies and a dance academy led by Rudolf von Laban. Notable residents and visitors, including Max Weber and Frieda von Richthofen, participated in cultural innovation that would echo globally. Still, this shift was not sufficient to guarantee economic sustainability. By 1920, Oedenkoven and Hoffmann left for Brazil to start anew. In 1926, the Monte Verità estate was sold to Baron Eduard von der Heydt, who transformed it into a private conference centre. Its transformation from utopia to a real estate commodity signals the limitations of charismatic cultural projects unsupported by inclusive governance or sustainable integration strategies.

In hindsight, Monte Verità's reliance on elite networks and symbolic appeal without substantial community outreach may be interpreted as an early form of soft power, i.e., projecting a radical image to attract like-minded participants and patrons. Yet its symbolic capital was insufficient to secure institutional or local legitimacy. The dilemma it faced between internal cohesion and external integration foreshadowed the contemporary challenges of Tamera.

Tamera is situated in a region facing severe ecological, demographic, and socioeconomic decline. Southwestern Alentejo is Portugal's largest and most sparsely populated region, comprising just 7.2% of the national population with only 23.7 inhabitants per km^2 in 2010 (Alves et al. 2015: 5). During the 20th century, the municipality of Odemira, where Tamera is located,

TABLE 2.1 Comparative dimensions of Monte Verità and Tamera

Dimension	*Monte Verità (1900–1926)*	*Tamera (1994–Present)*
Historical Context	Post-industrial anxiety, interwar Europe	Post-authoritarian democracy, EU expansion
Governance	Private ownership by founders	Owned by private company, steered by NGOs
Economic Strategy	Sanatorium, artistic retreat	Workshops, donor circle, ecotourism
Ideological Basis	Lebensreform, anarcho-spirituality	Post-patriarchal Eros, spiritual ecology
Community Outcome	Disbanded, absorbed by elite networks	Active, semi-integrated in local governance

underwent a process of natural and human desertification, which rendered it one of Europe's poorest and most climate-vulnerable areas (Esteves 2017: 980). In this context, Tamera's model of a post-capitalist community is both ambitious and culturally dissonant.

At the time of the fieldwork (2014–2015), approximately two-thirds of the 170 residents of Tamera were from German-speaking countries. The rest came from Spain, Israel, Italy, Britain, France, and the United States (US); only seven were Portuguese but were not born or raised locally. Tamera's internal economy favours those who can rely on personal wealth, rental income, or German fluency to secure seasonal work abroad. This enables partial withdrawal from the labour market and underwrites the time-intensive lifestyle of a cultural vanguard.

Tamera's economic model combines for-profit and non-profit streams, centred on education and global donor support. At the time of the fieldwork, over 60% of its income came from educational programmes. Additional funding came from a Support Circle, an international network of donors cultivated through workshops, personal contacts, and dedicated fundraising trips to Germany, the US, and Switzerland. This funding is organised by the Grace Foundation headquartered in Zurich, which states its mission as helping 'people with money, influence and other resources (...) invest in a new planetary culture'. The initial motivation behind Tamera's funding strategy was to pay off early loans and insulate the community from traditional banking systems.

In its early years, Tamera primarily used German for daily communication. While English became more common in public events and workshops, German remained dominant, which often required real-time translation. Several non-German speakers—especially Portuguese members—reported feelings of cultural exclusion. One called it 'a German bubble that is hard to burst' (local resident, fieldwork notes, October 2015).

Tamera's radical approach to sexuality, which is central to its pedagogy and public identity, also proved alienating to surrounding communities. At the time of the fieldwork, the 'Love School' and related programmes positioned sexual liberation as key to ecological and spiritual regeneration. These themes permeated not only workshops on relationships but also those on permaculture and renewable energy. But such integration of sexuality into broader sustainability discourse caused discomfort among many neighbours.

One Central European expatriate living nearby noted:

> That whole question of love and sexuality is a barrier. The people of the region are very conservative. We have a [local] neighbour who lives right next to them but doesn't even want to hear anyone talk about Tamera for that reason.
>
> *(Fieldnotes, October 2015)*

Tamera was often the subject of gossip in the nearby villages of Colos and Reliquias. Locals referred to its residents as 'promiscuous' and morally suspect. Among Portuguese environmentalists, mostly urban and from younger generations, Tamera remained largely unknown. Those who were aware of it described it as a culturally alien expatriate enclave:

> They have little to no interest in integrating into Portuguese culture.
>
> Their priorities reflect a level of privilege inaccessible to most of us.
>
> *(Informal conversations during fieldwork, November 2014)*

This illustrates the broader dilemma of cultural innovation: the effort to develop radically new ways of living can also reproduce new forms of social closure. Tamera's symbolic appeal—its soft power—attracts a global elite in search of transformation but limits its potential for meaningful inclusion within local or national frameworks. As happened to Monte Verità, the project risks becoming a world apart, sustained by external networks but estranged from its immediate social environment.

Tamera: Cultural mediation as prelude to institutional collaboration

Several Portuguese members of Tamera have assumed a central role in bridging with the population and culture of southwestern Alentejo, as well as the rest of Portugal. Between the mid-2000s and mid-2020s, this mediation paved the way for dialogue with the municipality of Odemira to co-create a regulatory framework for sustainable development in the region and to promote the local population's learning from the strategies of land management employed by Tamera.

The mediation included translating an earlier version of Tamera's webpage into Portuguese[2] and establishing relationships with journalists, documentary filmmakers, academic researchers, and political activists. It also included engaging the support of neighbouring villagers who, having lived and worked for many years in Central or Northern Europe, also became cultural mediators. These villagers include the owners of Café Marco in the neighbouring village of Reliquias, who had returned to Portugal after living in Germany and France for nearly two decades. Their business became the main dining and snacking spot for Tamera community members, which facilitated dialogue with the locals. A patron of Café Marco, who was the elected president of the Junta de Freguesia (the village-level municipal administration) in 2015, explained the role of former emigrants:

> (...) there are always conversations over beer or coffee, and people find they have interests in common. (...) We have Portuguese people, natives of this village, who lived in Germany or France. They invited them to their houses for tea. They have another understanding of their reality. (...) They and their guests also have a big impact in the local economy. They bring a lot of money to businesses in the area, cafés, restaurants, shops, the cooperative. They also order a lot of products from local farmers. (...) Thanks to these exchanges, people started opening up more. (...) A lady from Tamera now gives classes on medicinal herbs at the Cooperative. The local ladies come and also share their knowledge on medicinal teas and cooking herbs.
>
> *(Interview during fieldwork, April 2015)*

The same informant, who was an urban migrant for a long time before returning to his home village, also explained the role that public authorities play in this process of cultural mediation:

> Then I started taking groups of people there. It was my own initiative. Nowadays the Junta de Freguesia organises regular walks to Tamera. Thanks to these walks, the older folks started to realize that what they do in the Water Retention Landscape is not new for us. That's what we used to do here in the older times, the retention of rainwater in human-made puddles in the back yards. The older folks look at that and say 'That's what we used to do, that's what our parents did. Like with composting (...)'.
>
> *(Interview during fieldwork, April 2015)*

A core aspect of this mediation is the integration of Tamera community members into sports and cultural events organised by the villagers. According to a neighbour of Tamera:

> The biggest opening was about five years ago, thanks to the football matches we organised with the support of Junta de Freguesia. People

> invited them [Tamera residents] to take part in local amateur teams, and things developed from there. (…) We also have a yearly Tamera/Reliquias football match. (…) Now we have 'mixed' village parties, with folk music. The Tamera choir is invited to perform at those parties. The choir sings in the Church at Christmas. They carry the statues of the saints during the processions. Every Easter, we invite them to take part in the procession and they carry palm and olive branches. They always send people to take part in the organizing team of our regular village markets. They sell medicinal herbs, teas, and tree buds. The revenue contributes to the construction of the senior citizen's home here in the village.
>
> *(Interview during fieldwork, October 2015)*

Eventually, a partnership emerged between Tamera and local public authorities that was facilitated by the mediation efforts of Portuguese community members with insider knowledge of municipal administrative cultures. By the time of the fieldwork, Tamera members were participating actively in both municipal and borough assemblies in the municipality of Odemira. These engagements included deliberations on laws, policies, and joint initiatives aimed at poverty reduction and regional social cohesion.

This evolving relationship can be interpreted as a form of localised soft power: Tamera leveraged its symbolic and cultural capital not only to attract external supporters but to negotiate legitimacy within its host region. By presenting itself as a source of ecological, technological, and social innovation, Tamera gained a platform to influence public discourse and policy, despite its outsider status. The community's soft power rested on its ability to inspire rather than impose, offering an aspirational model of sustainability that appealed to municipal actors seeking alternatives to rural decline.

Regular open days and guided tours, offered in exchange for small donations, became key moments of cultural diplomacy for Tamera. These were complemented by workshops and talks conducted in Portuguese that focused on Tamera's research on social technologies, ecology, and community development. Through these activities, Tamera strategically bridged the symbolic gap between its countercultural ethos and the institutional logics of local governance, positioning itself not merely as a utopian enclave, but as a partner in regional resilience.

Conclusion

This chapter examines how green intentional communities that prioritise cultural innovation deploy post-patriarchal soft power as a strategy for survival, legitimacy, and influence. Through a comparative-historical analysis of Tamera and Monte Verità, it demonstrates that symbolic appeals to ecological regeneration, emotional transparency, and liberated sexuality serve as powerful tools of attraction, that enable these communities to mobilise support from privileged actors in mainstream society. This symbolic capital can offset the absence of public funding or formal regulatory frameworks conducive to commons-based governance. Yet it can also reproduce patterns of social selectivity and cultural exclusivity, which limit the potential for these communities to become truly inclusive models of sustainability transition.

The cases of Monte Verità and Tamera illuminate the paradox of countercultural soft power: while it enhances visibility and draws in resources, it tends to do so by attracting socioeconomically and culturally similar publics. This narrows the demographic base of participation and risks reinforcing the very hierarchies that such communities seek to overcome. In both communities, symbolic dissent from industrial modernity was grounded in the highly specific aesthetic, linguistic, and emotional repertoires of affluent, cosmopolitan Central Europeans. These repertoires have shaped not only the communities' internal 'lifeworlds', but also their modes of outreach, which can have a limited appeal for surrounding populations and underrepresented social groups.

Monte Verità's ultimate failure to integrate structurally or economically into its host context highlights the fragility of charismatic, culturally elite experiments when faced with geopolitical turbulence and material constraints. Tamera, by contrast, has sustained itself in part through its capacity to project cultural and technological soft power globally and locally. Its efforts at cultural mediation—particularly through the role of Portuguese members and former emigrants—offer important lessons in how countercultural communities can reframe their identity to engage meaningfully within their institutional and geographic contexts. However, these efforts still have not eliminated barriers to entry for those without sufficient cultural or economic capital.

Ultimately, this chapter argues that post-patriarchal soft power is both a resource and a constraint. It enables communities like Tamera and Monte Verità to attract symbolic and material support from a culturally aligned public. Yet without strategies for social translation and material inclusion, it can also reinforce elitism and symbolic closure. This is not a failure of vision but a structural limitation of relying primarily on aesthetic, emotional, and ethical charisma in the absence of broader institutional alliances or redistributive mechanisms.

A core insight from this chapter is that cultural innovation, while necessary, is not sufficient for long-term viability or inclusive transformation. Without structural strategies to redistribute access, whether through legal commons, shared ownership, or public subsidy, green intentional communities risk becoming exclusive enclaves that mirror rather than disrupt existing social divides. Even where ideological commitments to egalitarianism are strong, the operationalisation of these ideals can falter without mechanisms to ensure inclusion across class, nationality, and linguistic lines.

In order to move forward, green intentional communities must not only cultivate visionary lifeworlds but also adopt governance and funding strategies that support broader participation. This includes state-supported commons frameworks, sliding-scale participation models, and active co-design with local publics. Furthermore, comparative-historical research such as this should continue to track how charisma, privilege, and power circulate in utopian experiments—not to delegitimise them, but to make their transformative potential more accessible and socially embedded.

By foregrounding the tensions between post-patriarchal soft power and structural inclusion, this chapter contributes to the growing body of work that critically scrutinises the political economy of utopian projects. It also challenges both celebratory and dismissive narratives by showing how the aesthetics of transformation can be mobilised toward real social innovation, but only when anchored in frameworks that recognise intersectional power dynamics and actively redistribute access to such communities.

Acknowledgment

Ana Margarida Esteves' research was supported by national funds through FTC – Fundação para a Ciência e a Tecnologia (Foundation for Science and Technology) under the framework of UIDB/03122/2020 (DOI 10.54499/UIDB/03122/2020) and UIDP/03122/2020 (DOI 10.54499/UIDP/03122/2020).

Notes

1 The interlocutor quoted here gave permission to use her comment but chose to remain anonymous.
2 The current version of the website, launched in 2024, was produced in English. At the time of this writing, it has not yet been translated into Portuguese or German.

References

Altvater, E. (2005). Global order and nature. In R. Keil, D.V.J. Bell, P. Penz and L. Fawcett (Eds.). *Political ecology: Global and local* (pp. 19–44). Routledge.

Alves, D., Barreira, A., Carvalho, J., Carvalho, M., Cristóvão, A., Sequeira, T. N., & Fragoso, A. (2015). *Desertificação, desenvolvimento local e aprendizagem de adultos: Percepções dos actores locais no sudoeste alentejano* [*Desertification, local development, and adult learning: Perceptions of local actors in the southwestern Alentejo*]. Universidade de Lisboa.

Bourdieu, P. (1984). *Distinction: A social critique of the judgement of taste*. Harvard University Press.

Bourdieu, P. (1986). The forms of capital. In J. Richardson (Ed.), *Handbook of theory and research for the sociology of education* (pp. 241–258). Greenwood.

Chatterton, P., & Pickerill, J. (2010). Everyday activism and transitions towards post-capitalist worlds. *Transactions of the Institute of British Geographers*, 35(4), 475–490.

Duhm, D. (2010). *Eros unredeemed: The world power of sexuality*. Verlag Meiga.

Escobar, A. (2018). *Designs for the pluriverse: Radical interdependence, autonomy, and the making of worlds*. Duke University Press.

Esteves, A. M. (2017). Commoning at the borderland: Ecovillage development, socio-economic segregation and institutional mediation in southwestern Alentejo, Portugal. *Journal of Political Ecology*, 24(1), 968–991.

Federici, S. (2018). *Re-enchanting the world: Feminism and the politics of the commons*. PM Press.

Forno, F., & Wahlen, S. (2022). Environmental activism in everyday life. In M. Giugni & M. Grasso (Eds.), *The Routledge handbook of environmental movements* (pp. 253–266). Routledge.

Foucault, M. (1977). *Discipline and punish: The birth of the prison*. Pantheon Books.

Fraser, N. (1997). *Justice interruptus: Critical reflections on the 'postsocialist' condition*. Routledge.

Gibson-Graham, J. K., Cameron, J., & Healy, S. (2013). *Take back the economy: An ethical guide for transforming our communities*. University of Minnesota Press.

Gross, O. (1913). *Die freie Ehe* [*The free marriage*]. Verlag der Zukunft.

Koselleck, R. (2004). *Futures past: On the semantics of historical time*. Columbia University Press.

Kunze, I. (2012). Social innovations for communal and ecological living: Lessons from sustainability research and observations in intentional communities. In A. Franz-Balsen & M. Hirschnitz-Garbers (Eds.), *Sustainability in science and practice* (pp. 189–206). Oekom Verlag.

Landauer, G. (2009). *Revolution and other writings: A political reader*. G. L. Lyman, Ed. & Trans. PM Press.

Lichtenfels, S. (2000). *Traumsteine. Reise in das zeitalter der sinnlichen erfüllung Dreamstones.* [*Journey into the age of sensual fulfilment*]. Verlag Meiga.

Marcus, G. E. (1995). Ethnography in/of the world system: The emergence of multi-sited ethnography. *Annual Review of Anthropology*, 24, 95–117.

Meijering, L., Huigen, P. P. P., & van Hoven, B. (2007). Intentional communities in rural spaces. *Tijdschrift voor economische en sociale geografie* [*Journal of Economic and Social Geography*], 98(1), 42–52.

Melucci, A. (1996). *Challenging codes: Collective action in the information age*. Cambridge University Press.

Mongini, N., Gatti, C., & Risaliti, S. (Eds.). (2023). *Monte Verità: Back to nature*. Edizioni Lindau.

Mühsam, E. (1999). *Unpolitische erinnerungen* [*Unpolitical memoirs*]. Karin Kramer Verlag.

Mühsam, E. (2002). *Fanal: Revolutionäre gedanken und gedichte* [*Beacon: Revolutionary thoughts and poems*]. Karin Kramer Verlag.

Mühsam, E. (2002a). *Der Bolschewismus und die anarchie* [*Bolshevism and anarchy*]. Karin Kramer Verlag.

Nye, J. S. (2017). Soft power: The origins and political progress of a concept. *Palgrave Communications*, 3, Article number 17008. https://doi.org/10.1057/palcomms.2017.8

Pickerill, J., & Maxey, L. (2009). Geographies of sustainability: Low impact developments and radical spaces of innovation. *Geography Compass*, 3(4), 1515–1539.

Pinto, T., & Vilaça, H. (2023). New Age and environment: New forms of spirituality and lifestyle in the context of secularization? *Religions*, 14(4), 468.

Russo, D. (2024). Eros and ecology – The new alchemy of sustainable design. *Design for Inclusion*, 128, 145–150.

Scott, J. C. (2009). *The art of not being governed: An anarchist history of upland Southeast Asia*. Yale University Press.

Seyfang, G., Smith, A., & Longhurst, N. (2010). Grassroots innovations for sustainable development: Towards a new research and policy agenda. *Environmental Politics*, 19(4), 564–582.

Sirost, O. (2016). Les Kosmiker [The Cosmics]. *Sociétés*, 134(4), 29–39.

Tollefson, J. W. (1991). *Planning language, planning inequality: Language policy in the community*. Longman.

Turner, F. (2006). *From counterculture to cyberculture: Stewart Brand, the Whole Earth Network, and the rise of digital utopianism*. University of Chicago Press.

Walker, G., & Devine-Wright, P. (2008). Community renewable energy: What should it mean? *Energy Policy*, 36(2), 497–500.

Weber, M. (1978). *Economy and society: An outline of interpretive sociology*. G. Roth & C. Wittich, Eds. University of California Press.

Wolsink, M. (2020). The research agenda on social acceptance of distributed generation in smart grids: Renewables as common pool resources. *Renewable and Sustainable Energy Reviews*, 127, 822–835.

3

MAJORITY-RULE, SOCIOCRACY, AND DO-OCRACY

Micropolitics in Danish green communities

Quentin Gausset, Camilla Nielsen-Englyst

Introduction

Green communities are groups of people who join forces collectively to live more sustainably or to live in a more sustainable environment (Gausset 2020; Gausset & Hoff 2024). Collective action requires coordination and can lead to different kinds of social organisations (different forms of green communities) that are characterised by different ways of making decisions (different forms of micropolitics managing time, responsibilities, and/or resources). Our chapter introduces sociocracy and do-ocracy as emerging alternative forms of decision making, discusses when and where these forms of micropolitics are used, and how they influence social interaction within Danish green communities. We explicitly focus on green communities that are *practivist*, in the sense that practivist people focus on changing their own practices—in contrast to activists focusing on changing other peoples' practices (Hansen 2019; Hoff, Gausset, & Lex 2020).

The three green communities discussed in this paper, an ecovillage, a green neighbourhood and a bee-keeper guild, have very different goals and organisation. Although they are not representative of Danish green communities in general or of ecovillages, green neighbourhoods or bee-keeper guilds in particular (as the variation within each of these categories is too big to claim that one example can be representative), they face the same type of challenges regarding decision making mechanisms. They also all rely on different forms of sociocracy and do-ocracy in order to avoid some of the pitfalls that characterise majority rule and consensus.

The empirical material presented here draws from first-hand social interactions and observations. Since 2022, the first author has conducted

DOI: 10.4324/9781003769064-4

participant observation and engaged in research in three different green neighbourhoods (green communities in which neighbours meet and work together around collective green activities) and participated in more than 40 communal meals, 20 thematic activities of different working groups, six green high schools, and many informal conversations with participants at each of these events. He has also been involved in all the meetings of the 'green brain', a sub-group of participants that has defined the vision, values, and organisation of the National Association of Green Neighbourhoods in 2022–2023.

The second author has lived for 20 years in an ecovillage that she was part of forming and has been part of launching bofællesskab.dk in 2018, an association supporting the development of cohousing in Denmark. She has been the vice chairperson of the Danish Ecovillage Network, served five years on its board, and is currently employed in that organisation as a project leader. She also works for a developer that launched its first cohousing project in June 2025. Through long-term observant participation and personal social engagement, both authors have accumulated first-hand knowledge of the green communities described in this chapter, including both their daily activities and the overall strategic organisation of the broader movements in which these green communities are inscribed.

Associations and communities: Majority rule and consensus

Denmark takes pride in being a country of associations (*foreninger*). Voluntary associations are widely considered one of the backbones of grassroots democracy in the country (Clemmensen 2008; Henriksen 2004; Gundelach 1984; Gundelach & Torpe 1997). One finds thousands of associations located everywhere in Denmark, including housing associations (cohousing associations and ecovillages), street owner associations, sports associations, school associations, and associations working for human rights, social issues, climate, and biodiversity issues. All associations are organised around rather standardised formal constitutions that specify majority rule, board members elected during general assemblies, and other bureaucratic rules requested by the authorities, including banks or municipalities, that register associations and interact with them.

But along with these formal associations, one also finds in Denmark many informal communities (*fællesskaber*) that are much more loosely organised: they have no board members or majority rule and are not officially recognised by the state, even though they can receive support from municipalities, for example, when borrowing a municipal room for their activities. They are simply groups of people who meet around a common interest, organise themselves informally, and keep meeting until circumstances change and the community disappears. Examples range from groups of friends who meet at regular intervals for social activities to social clubs that meet informally around a common activity (knitting, jogging, and the like). Groups of neighbours who meet to discuss how they can live more sustainably are referred to here as 'green neighbourhoods'. While there are an estimated 100,000 civil society organisations in Denmark (i.e., one per 60 inhabitants), including associations, foundations, and self-owned institutions (Selle et al. 2019: 48), we have no way of measuring the number of informal communities. Although these informal communities are everywhere, they tend to operate under the media and bureaucratic radar as they are ephemeral and informal and remain mostly unaccounted for.

Informal communities (sometimes also called 'non-membership-based networks') are growing in popularity because they require only informal engagement and low commitment. At the same time, formal associations are losing some of their former popularity because people tend to shy away from bureaucratic organisations, and it is becoming increasingly difficult to recruit association board members (Torpe 2003; Grubb & Henriksen 2009; Højgaard 2020). While engagement in a formal association is rather 'obligation-driven' and requires a stronger commitment, engagement in informal communities is comparatively mostly 'lust-driven' and based on personal passion or interest. This factor might help explain the rapid spread of the green community movement in Denmark (Nielsen-Englyst & Gausset 2024). And yet, despite their popularity and although they play an essential role in the daily social fabric (Gausset & Hoff 2024), informal communities are much less studied than formal associations unless they are organised under an umbrella national association that advertises and represents their interests.

As a rule, associations rely heavily on majority rule, while communities prefer to reach decisions through consensus, even though associations can also rely on consensus, for example, by preferring to make a decision by acclamation rather than by a formal vote. Communities may be compelled to become formal associations and adopt majority rule whenever they grow beyond a certain threshold and/or manage significant common resources, as exemplified in the three case studies discussed below.

Majority rule has the advantage of being a 'fast' collective decision making mechanism, one in which a majority imposes its view on a minority. Decision making is even faster in representative democracy when majority rule is used to elect board members who are given ample room to make decisions for the group, which reinforces the marginalisation of minority views. By contrast, making decisions based on consensus protects minority views, but the process is time-consuming and the veto right conferred on every member can slow down the decision making process and sometimes block it entirely.

Although majority rule and consensus are defined in opposition to minority rule (also called tyranny when it is reduced to one person's decision), both can be hijacked by individuals who can succeed in imposing their personal view on the group, either by blocking decision making in consensus rule or by monopolising positions of powers that no one else opposes when such an individual runs for re-election. Concrete empirical examples of both problems abound, including in green associations and communities. Moreover, both majority rule and consensus often split the group into those supporting a decision and those opposing it, which can threaten group cohesion.

This chapter discusses how sociocracy and do-ocracy constitute alternative forms of micropolitics used in Danish green communities to create an agile decision making mechanism that avoids both minority rule (the hijacking of decision making by one or a few individuals) and splitting the community.

Sociocracy and do-ocracy as agile and inclusive forms of micropolitics

Sociocracy is a term coined by Auguste Comte (1851) that referred originally to a form of government run by sociologists ('experts' in society) making decisions on the basis of science and reason. While Lester Frank Ward (1883) tried to popularise Comte's ideas in the 19th century, it was only in the 20th century that sociocracy started being used as a new form of governance in Dutch Quaker circles when Kees Boecke established a residential school in 1926 (Priest et al. 2024). Sociocracy took on a new life in the 1970s, when Gerard Endenburg, a former student in Boeke's school, developed and formalised the 'sociocratic method' as a form of decision making that aims to avoid the pitfalls of majority-rule, minority-rule, and pure consensus by balancing efficient and egalitarian decision making (Endenburg 1974; 1975; 1998; Buck & Villiness 2017). Sociocracy has become a popular form of governance in eco-communities, not just in Denmark but also elsewhere (Bertrand 2022; Christian 2014; 2016; Dietvorst 2024; Rios 2011; Straub 2021; Tabor 2024; Tijhuis 2021; Yang 2021). It is also increasingly popular as a form of 'lean' management in business (Czekaj, Walczak, & Ziębicki 2020).

In a sociocratic decision making process, a project is proposed by one or several members to the group. Group members sit in a circle, and a facilitator gives everyone in turn the opportunity to raise legitimate concerns (appealing to reason rather than emotion or personal opinion), and/or suggest how the project can be improved. Each participant can also bring 'paramount objections' if they feel that the proposal, if passed, would have devastating consequences for themselves or for the group. Paramount objections and legitimate concerns trigger a new round of discussion in which everyone has the opportunity to suggest ideas to improve the proposal. As long as one member keeps having a paramount objection, the project cannot be approved (see also Owen & Buck 2020).

Although this sounds like a consensual form of decision making, sociocracy is fundamentally different in that it only accepts reasonable concerns based on rational arguments, thereby following Comte's original idea. Sociocracy thus rejects as illegitimate those concerns that are based on personal taste, opinion, emotion, conflict with other decision makers, or other aspects that do not directly relate to the topic discussed and the shared purpose of the group. In practice, however, the rejection of personal concerns is rarely done overtly. Instead, there is a strong focus on encouraging participants who come with emotional concerns to transform them into rational arguments that are linked to the purpose of the group and that can be incorporated into the decision making mechanism. In other words, sociocracy 'shifts supremacy from people to sound arguments' as 'people—regardless of their position, rank, function or role—are unable to block decisions based solely on opinion, personal preference or rank' (Priest et al. 2024: 33).

To function, sociocratic principles need to be understood and accepted by all participants, which requires continuous training, the onboarding of new members, and periodic review of roles. Many conflicts in intentional communities are fundamentally about preferences, values, or emotional safety. Yet, because sociocracy formally excludes personal preferences, emotions, or interpersonal conflicts as valid grounds for objection, it may sideline people unable to translate emotional discomfort into purpose-aligned 'rational' arguments, which can create subtle hierarchies privileging those with verbal and analytical competencies. And in the context of Danish intentional communities, which have historically shown a strong preference for pure consensus, the fact that some participants may understand sociocracy as a process intended to produce unanimous consensus rather than consent ('good enough for now, safe enough to try') can diminish the efficiency that sociocracy is meant to provide and can reproduce the dysfunctions that sociocracy is meant to overcome.

Notwithstanding these words of caution, sociocracy is practiced in many Danish ecovillages that see it as a more constructive and inclusive approach than pure formal consensus. It saves time in that it does not require everybody to form an opinion and agree on everything, and it leads to the adoption of flexible and improved projects rather than the adoption or rejection of rigid and immature projects. People are not asked to approve a project; they are asked to consent to it and to be satisfied with a 'I can live with it without leaving the group' perspective. All members must be willing to understand other people's legitimate concerns, make room for them, and let their personal egos vanish for the sake of the community.

Do-ocracy is another alternative to majority rule and consensus. It focuses on 'doing' rather than 'discussing and agreeing'. In do-ocracy, those who want to do something together either empower themselves or are empowered by a larger group to make decisions on how the project should be done. In practice, those who have an idea and want to do something just go ahead with their project together with those who just want to be part of it. Everybody is welcome to participate, but non-participants have no direct influence on how things are done. The term has been popularised by the transition town movement in the United Kingdom (UK) (Simmonds 2016) and by the Occupy movement in the United States (US) (Barnette & Barrett 2015). In Denmark, the term is mainly used in eco-communities and in the climate movement.

Although one does find manuals about how to organise do-ocracy (e.g. Simmonds 2016), planning for it seems more like a contradiction in terms because do-ocracy is most often purely informal and something that happens by itself because it works pragmatically without being explicitly planned. Likewise, presenting do-ocracy as a trend in formal participatory policies appears to be another contradictory statement, as it is a form of active citizenship that emerges from the grassroots rather than from the top (Blijleven 2016). The fact that some governmental organisations are beginning to acknowledge its worth and promote it (at least when it fits bureaucratic goals), as is the case in the Netherlands, is a positive sign, but do-ocracy can be problematic if its institutionalisation spoils its grassroots nature (Vrielink, Verhoeven, & Wijdeven 2013).

On the downside, the informality of do-ocracy can raise questions regarding its accountability (Blijleven 2016). Depending on the scale of the organisation, it can lead to 'creative chaos', but it can also end up as pure chaos if there is not a minimum of organisation and structure (Chen 2009). Moreover, it can lead to a 'tyranny of the pro-active' and create its own power imbalance because a small group can end up doing something without discussing it with the larger community and presenting the end result as a 'fait accompli' (Zarinah 2016).

Even in a do-ocracy, some people must take the lead while others become followers. Finding the right balance between leadership and inclusion is a classic challenge for all do-ocratic endeavours. If participants feel disempowered, they will sooner or later leave the group, the dominating or self-appointed leader will be alone, and this will signify the end of the group or the project. If participants immediately agree on what to do and how to run the project, they just go ahead. Those who do not buy into the project or its organisational setting can just stay home or engage in their own parallel alternative project. The informality of do-ocracy is therefore both its main strength and its main weakness. If it succeeds in balancing freedom and inclusion, do-ocracy is a fast, unbureaucratic, and efficient form of decision making, but one that relies on decentralising decision making in sub-groups that share the same vision and motivation.

Sociocracy and do-ocracy are alternative kinds of decision making that prevent the concentration of power in the hands of a few while at the same time securing a place for agility in decision making and avoiding the splitting up of the community. Sociocracy does this by formalising equal rights and by giving pre-eminence to reason and community building, while do-ocracy achieves this by decentralising decision making in smaller groups that take their independence from the control of the larger group.

Both sociocracy and do-ocracy emerge from the wish to mitigate the weaknesses attached to consensus while retaining its strength. The process of reaching consensus in sociocracy is negotiated and avoids veto rights. In do-ocracy, consensus is reached pragmatically without being discussed much. Thus, instead of being based on pure and formal consensus in which everyone has the right to veto, sociocracy is a form of reasoned and negotiated consent making with no room for personal emotions or opinions, while do-ocracy is a form of practical and organic consent in which initiative-takers cannot impose their will on followers.

In the next section, we discuss how sociocratic and do-ocratic decision making unfold in three Danish green communities.

Empirical cases

Case 1: Hallingelille ecovillage

Hallingelille is an ecovillage established in 2005 and comprises 25 households (55 adults, 35 children). Like most Danish ecovillages, it is organised formally as an association with a normal constitution based on majority rule. This is an administrative necessity in order to buy land, secure loans, and interact with the municipality. Living in an ecovillage is a big investment in terms of time and money, and having a formal constitution provides some form of security for this investment. But one thing is what is written down and used for external recognition, another is everyday practice. When it comes to organisation and decision making, majority rule is most often avoided and more inclusive patterns are widely preferred, such as sociocracy in community meetings or do-ocracy in working groups.

The Hallingelille constitution states that consensus is preferred, but that voting can take place as a last-case scenario (voting has happened only a few times in 20 years). In the early days of the ecovillage, consensus was preferred for decisions, and an agenda point could go through or not, depending on consensus and vetoes (a win-or-lose scenario). 'Did the point go through?' could be a question asked by a community member who did not attend a meeting. After several experiments, Hallingelille's decision making mechanisms were transferred in 2017 to two different decision making bodies: (1) common meetings, in which sociocracy is used to facilitate decision making, and (2) working groups, to which common meetings delegate power to decide on certain areas and themes and which have received since then increasingly more precise domain descriptions with dedicated budgets.

Hallingelille attempted to integrate sociocracy into its constitution, but this was rejected by its bank and municipality. As a result, the residents decided to use sociocratic principles informally and pragmatically without having the process backed by the support of the community's constitution. A general assembly is organised annually in accordance with the constitution, but it is conducted as a brief portion of a regular common meeting and is expedited as a formality, simply to fulfil bureaucratic obligations.

The use of sociocracy in Hallingelille from 2017 on represented a major shift from both majority rule and pure consensus-based decision making. The common meetings moved away from a rigid, all-or-nothing (winner or loser) scenario in which a few individuals could block decision making and moved towards a truly collective and fluid process in which a project might end up being very different from the initial proposal but supported by everyone.

An example of this process occurred when some residents proposed having two pigs that would occupy the shared communal fields. As the point was being discussed with sociocratic facilitation, those objecting to the initial proposal suggested what could be a better decision for the community that involved both the number of pigs and their location. This sociocratic exercise demanded a high level of trust among participants, i.e., trusting that every member would act in good faith to find the best solution and trusting that the best solution would eventually be reached, even though the outcome was not precisely known at the time the decision was made.

With time, as trust became established in the community and the sociocratic processes matured, more and more autonomy was delegated to the working groups. Today, a new project such as 'pigs in the shared area' would not even be discussed as an agenda item at the common meeting. It would simply be delegated to the group taking care of the common facilities (as part of their approved domain) that could be trusted to make the best decisions in their area of expertise.

Adopting sociocracy has led to more efficient and more agile decision making based on reasoning rather than emotions. An example of this is when the question of insurance for shared facilities was raised in Hallingelille early in the sociocratic development process. At that time, one person objected to insuring the shared facilities because he disliked insurance companies. When the community carefully explored the objection from a reason-based and not emotionally charged perspective and related the issue to shared values and the purpose of the community, the insurance proposal was not blocked and abandoned because of the objection of one participant. Additional language was included that insurance should be managed by a value-based and user-owned insurance company. This decision not only mitigated the objection, it also caused the community to make a decision that was closer to its values, which resulted in a better decision by integrating the objection. Since emotions are discouraged in the operational decision making forum, a separate meeting forum has been created to give members the opportunity to express and discuss their feelings. The room in which decisions are made and the room in which emotions are expressed are kept strictly separate.

Hallingelille now has 27 different working groups that work in five areas of community life: indoor facilities, outdoor facilities, social activities, administration, and development/innovation. These groups all operate under approved domains and budgets (decided for a two-year periods) and make independent decisions about their domain of responsibility. This decentralisation of decision making means that the ecovillage is no longer a structure in which 'everybody decides everything'; it is a place where those with the most knowledge and interest decide. Those closest to the impact of the decision have a higher say that is inspired by the 'advice process' (Laloux 2014). In other words, this type of decision making involves those who would be most affected by a project rather than forcing everyone in the community to have an opinion about everything.

In parallel to the two formal decision making bodies, residents are free to come up with any project they have in mind, both individually and collectively. Someone might wish to build a nest for storks or a hedgerow fence and invites anyone interested to join the project. Those who join then decide collectively and do-ocratically how it should be done without any interference from non-participants and with a strong focus on doing things rather than discussing them.

Some ecovillagers who believe in pure consensus in which it is important that everyone agrees on everything express frustration that fewer people show up at the common meetings today than before. Others are pleased to delegate decisions to those who know the most (the working groups) to reserve the common meetings for vision making and substantial decisions, thus avoiding discussions of trivial questions that can be extremely time-consuming. Most people seem to enjoy working together on a specific project rather than participating in numerous meetings.

This case shows how this hybrid governance in Hallingelille mixes different forms of micropolitics to maximise efficiency and community cohesion. While the ecovillage was first forced to adopt majority rule in its constitution, it chose in its early years to sideline this form of micropolitics to adopt pure consensual decision making. After experiencing the drawbacks of consensus, the ecovillage decided to adopt a combination of sociocracy and do-ocracy, with do-ocracy becoming increasingly important over time. This strategy has freed up time for members because the long agendas and endless common meetings of the early years have been considerably reduced, and the working groups are operating increasingly more autonomously. Majority rule can still be used when what is at stake economically or strategically is too controversial to find consent for any given position. Nevertheless, relying on majority rule is experienced as a communal failure to reach the consent strived for; it is seen as a last resort when everything else fails.

Case 2: A green neighbourhood in the Copenhagen area

Green neighbourhoods are groups of people living in the same neighbourhood who meet around common activities that promote more sustainable living. The movement started in 2020 and today totals more than 360 local green neighbourhoods and more than 56,000 members.

The green neighbourhood in this case was launched in 2022. Its main activity is a meat-free and organic common meal in which 15 to 35 people meet once per month. After dinner, people discuss the topic of the day or plan collective activities. With time, other activities have been added to the communal meal. There is a garden group in which members visit each other's homes to learn new ideas for more biodiverse and climate-friendly gardens. These ideas include, for example, drought-resistant plants, insect-friendly flowers, dead hedges or trees that store carbon, water ponds, homemade compost, and so on.

This green neighbourhood also organises a swap market once per year in which up to 200 people from the neighbourhood and beyond come to get rid of the clothes, toys, books, kitchen utensils they no longer use and/or find something that they can use and take home, for free. Moreover, nine members have started to order a vegetable box from a regenerative agriculture farm with 14 other households who have joined this sub-group even though they do not participate in any other activity of the green neighbourhood. The vegetables are brought in bulk once a week to a collecting place, and subscribers come and choose their vegetables at this local collection spot.

Another collective activity of the green neighbourhood pertains to energy consumption at home. Some members meet in private homes to talk about house insulation, energy savings, solar panels, rainwater catching, and other housing-related sustainability issues. In yet another sub-group, people meet in private homes to discuss upcycling workshops, such as braiding mats with old clothes, making homemade soap and beauty products, or discussing how to adopt a zero-waste lifestyle. The neighbourhood has also a system of 'neighbour-car': three car owners lend their cars to seven families who do not own a car; the families share maintenance costs each time they borrow one.

While the common meal remains the most frequent activity and the one with the most regular attendance, it is only one of seven main subgroups of activities; although many participate in several sub-groups, nobody participates in all activities. About ¼ of the car-sharing sub-group, ⅓ of the garden sub-group, ⅔ of the people taking a vegetable box every week, and ¾ of the people participating in the annual swap market do not participate in any other activity of the green community (see Figure 3.1). In fact, they often do not even perceive themselves as being members of the green neighbourhood. They are just happy to participate in the specific activity, either under

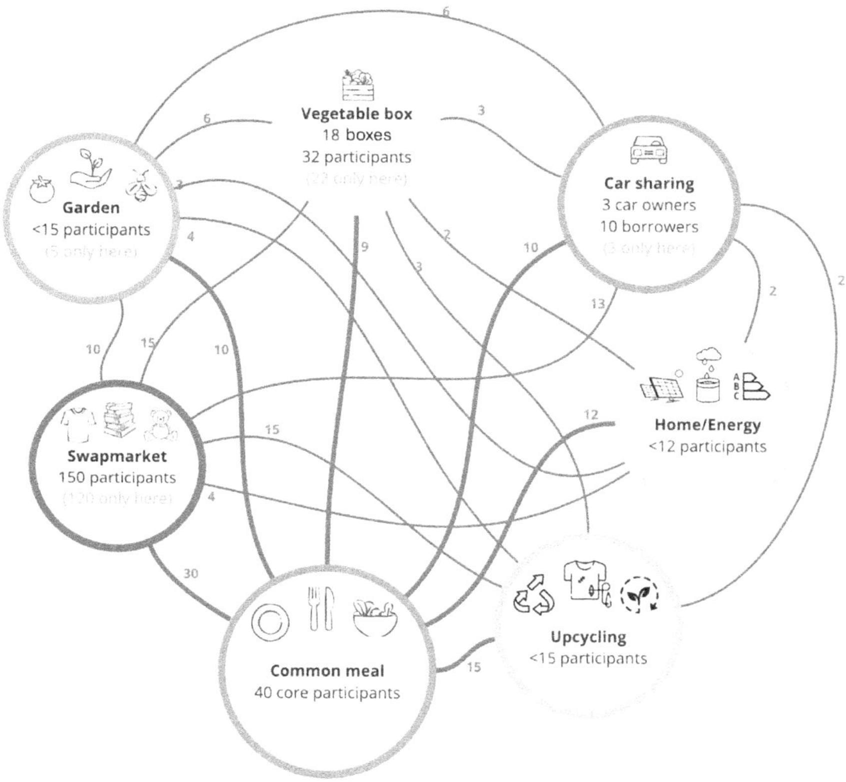

FIGURE 3.1 Map showing the seven neighbourhood sub-groups, the number of participants in each sub-group (+ eventually the number of people participating only in that specific sub-group), and the number of people participating in several sub-groups (numbers on each line)

another flag (such as the associations 'practical ecology' or 'green churches' in an informal collaboration with the local green neighbourhood) or as an individual without any collective affiliation. Thus, in practice, the green neighbourhood is not a coherent community in which each member knows everybody else. Instead, it is a collection of seven decentralised thematic sub-groups dealing with specific aspects of sustainability as depicted in Figure 3.1.

This local green neighbourhood is not an association: it has no constitution, no elected board, no general assemblies, no membership fee, and no bank account. It is entirely driven by internal motivation and ad hoc contributions, when needed, and everybody is welcome. Someone who has an idea just sends an invitation to others to create a collective activity. Those who are interested and have time come to the first group meeting and things evolve from there in an informal manner.

The common meals and swap markets require more organisation than the other activities: a small core group takes the lead to book a room, find cooks and volunteers, send invitations, and collect money if necessary. Other sub-groups are more loosely organised: someone who would like to show or discuss something invites others to his or her home, or someone expresses interest in something that someone else can demonstrate (how to make compost, how a solar panel system works, etc.), which leads to a collective invitation. In the best cases, someone takes the lead to plan and organise a succession of meetings in different private homes; in other cases, invitations are somewhat improvised. Each sub-group has its own pool of initiative-takers, although there is some overlap between sub-groups. The vegetable box sub-group is mainly organised by the regenerative farm and the neighbour-car sub-group is highly decentralised, with borrowers contacting lenders directly whenever they need to use a car.

A green neighbourhood functions only because some people take the lead to invite others to an activity and because some respond to the invitation and participate actively. The model is not one of an association in which a core elected group organises services and activities for others, but rather that of a community in which most members are actively co-organising and taking part in the activities—if they do not, the community simply vanishes. There is no formal seat of power: it is the people who take the initiative to invite others who set the agenda, but the people who attend do have a say in how the activity unfolds. The daily micropolitics is therefore do-ocratic, with a focus on doing things together without wasting time discussing how they should be done, even if this leads to a certain degree of improvisation.

The local green neighbourhood communicates through a mailing list containing the addresses of every person who participated in at least one activity and who voluntarily left his/her contact details (roughly 100 in total). It also has a Facebook page with more than 130 members. Only 35% of these Facebook members are also present on the mailing list, while most of them (around 85) have never participated in any physical activity and are either 'passive' members (around 70) or members of other green neighbourhoods who are curious to see the activities organised by this one in particular (around 15). If one only considers those who participate regularly in more than one collective activity, the number is closer to 40 households in a neighbourhood that has between 700 and 800 households, which is about 5% of the total.

The Facebook page of the group was created by the National Association of Green Neighbourhoods, which works as an umbrella structure for scattered green neighbourhoods and posts messages of general relevance on the local Facebook pages. The association also raises funds to help increase the number of new green neighbourhoods and to support existing ones. It regularly organises green 'high schools' to inspire and help those taking the initiative in local green neighbourhoods and to facilitate networking among them. It contributes to organising a yearly National Climate Action Day and it makes green neighbourhoods known to municipalities, policymakers, and the national media. It also allows local green neighbourhoods to use its CVR number (the central business registration number) to apply for project funds even if they do not have local constitutions. The National Association of Green Neighbourhoods has a registered constitution, a bank account, a CVR number, and a board that is elected by representatives of local neighbourhoods in a yearly general assembly.

Local green neighbourhoods are communities that can afford to work informally and do-ocratically because they do not share common resources and because there is much less at stake than in an ecovillage. The loose networks of neighbours who meet to engage in collective green activities are often not even aware that they are part of a national movement or that there is a national secretariat that can support their activities. By contrast, the national organisation representing green neighbourhoods is an association with a formal constitution, the protocols of representative democracy, and majority rule, all conditions required by municipalities and donors alike.

Case 3: A bee-keeper guild

The local guild (*laug*) of bee-keepers described in this case was founded in 2012 by eight people who did not know much about bee-keeping and did not know each other. But they wanted to keep bees and promote knowledge about them and their role in nature. The group applied for and received money from the local municipal committee (*lokaludvalg*) to buy bee equipment (beehives, a manual honey extractor, and so on). They were granted the money under the explicit condition that they be formally organised as a registered association with a constitution based on formal majority rule.

Since the launching of the guild, four new members have joined the group, and more than half of the original members temporarily stopped keeping bees at some point, which means that those who participate in the regular meetings usually number no more than ten. Each active bee-keeper takes turns to invite other members at home for a cup of coffee and to touch base about how everyone's bees are doing. Once a year, after an hour of the usual coffee and discussion, one of these informal meetings is formally called the yearly general assembly. The meeting then takes the standard form of any Danish association: agreeing on the agenda, appointing a moderator, appointing a rapporteur, hearing the yearly account from the head of the board, the financial account from the treasurer, and voting for new board members (headman and cashier). The financial account is quickly expedited: there has never been any guild membership fee, there are basically no expenses, and for many years the total assets of around 40 euros have been kept in a cigar box.

The guild is officially an association with a formal constitution, but it functions more as a loose community. The organisation of the guild has repeatedly been the object of internal debate: should it continue to be a formal association or should it simply be a community (a 'come-together group', as expressed by one of its members) since all activities are pretty much informal except when they are artificially formalised once per year during the general assembly.

The usual informal routines took another turn the day someone declared in a yearly general assembly that he was considering buying an electric honey extractor and asked if anyone was interested in sharing the cost with him. He also suggested that if everyone was interested, it could be bought by the guild and partly paid for by selling the manual honey extractor that no one had been using for several years. (The manual extractor was too hard to operate manually and most members either pressed their honey with an apple press or borrowed an electric honey extractor from someone else). Some, but not all members present were interested in buying an electric honey extractor. They quickly decided to buy one that was on sale in an online shop and agreed to sort out the details of the collective payment and responsibilities later.

This decision stirred a heated debate about how the guild should manage the details of this issue. First, those who were not present during the general assembly felt that the discussion of this issue should have been announced formally beforehand (as stated in the association's official constitution) and called for an extraordinary general assembly with this point on the agenda. Second, the debate that ensued discussed two models, including one in which a few members were to buy the extractor on their own. This option presented a difficulty: would others be granted access to it and under what conditions? Would that split the guild in two distinct sub-groups? The other option would have the association finance it, but in that case should everyone contribute to the purchase, even those who would never use the extractor?

As a relative consensus seemed to emerge for the second option, the debate took on a third dimension about how to finance the purchase: by the introduction of a yearly membership fee or on the basis of differentiated voluntary donations or through a one-time contribution that would be the same for all? This last option was finally chosen, and every member (even those who were no longer keeping bees or those who used their own apple press to extract honey) agreed to contribute with a one-time payment of 45 euros which, together with the money generated from selling the manual extractor, covered the cost of the electric one. The meeting ended in general relief that a consensual solution could be found without having to resort to a vote that would have split the group by pitting some members against others. Some members concluded the meeting by proudly seeing it as an exemplary case of direct democracy.

This case shows how do-ocracy is an advantageous form of micropolitics in common meetings, but also how do-ocracy must give way to either formal majority rule (as when the guild applied for financial support from the local municipal committee) or to a sociocratic exercise (as when a new motorised extractor was purchased) as soon as significant collective resources come into the picture.

In other words, the new extractor introduces a cost for members and transforms the usual non-monetary collective dynamic of the group. Conflicting interests threaten to split the guild. Bureaucratic association statutes are called upon to prevent this from happening, but contrary to the constitution, a vote is avoided at all cost in order to find a sociocratic solution that takes everyone's concerns into account (including those who temporarily stopped keeping bees) and in which everyone consents to a series of step-by-step decisions for the sake of moving forward while keeping the group together.

Discussion

Although very different in some ways, these three cases show striking similarities in the way majority rule (usually referred to as 'democracy' in Denmark) and more consensual ways of making decisions (which include pure consensus, sociocracy, and do-ocracy) unfold in practice.

First, micropolitics in green communities is most often hybrid, mixing majority rule, pure consensus, sociocracy, and do-ocracy in complex ways. Regardless of whether communities have constitutions or not and regardless of what is written in constitutions, all communities mix different forms of micropolitics, consciously or not, and try to find the one that that provides time-efficient decision making while avoiding conflicts and frustrations within the community.

Second, micropolitics is influenced by the question of scale. Consensual decision making is easier to achieve in smaller groups; all green communities are compelled to organise in formal associations when they reach a certain size and scale and start managing or applying for funds. An associative structure with a formal constitution and majority rule is the only model that is seen as legitimate by municipalities, funds, and banks.

Thus, we can observe a dual micropolitical field: when green communities interact with official authorities and institutions, they put on the clothes of majority rule and representative democracy. But when they interact locally, green communities try to reach consensual decisions in their everyday political practices to avoid dividing the community, despite having constitutions saying that decision making is conducted on the basis of majority rule. We see in all these cases (and in most Danish green communities, generally speaking) a strong will to bypass majority rule whenever possible and to rely on shadow consensual decision making, either formally facilitated by a moderator according to a set of proven methods or informally negotiated without moderators or through do-ocracy. Thus, in practice, the search for consensus remains the ideal; majority rule is only used as a safety net if it is impossible to reach an agreement through more consensual and less divisive decision making processes (such as sociocracy and/or do-ocracy). Relying on majority rule is often seen by community members as a collective failure and may sometimes be seen as likely to have deep and long-lasting consequences for community cohesion.

Third, the balance between majority rule and consensual decision making (including sociocracy and do-ocracy) depends on the importance given to inclusion and community cohesion as opposed to fast and efficient decision making. Both majority rule and consensus risk dividing the community into winners and losers. In this regard, sociocracy and do-ocracy represent agile ways to make decisions while at the same time preserving community cohesion.

Fourth, the choice of micropolitics is linked to the need to find the right balance between community control and personal freedom, which is a crucial issue for the survival of any green community (Hansen 2020). Both an excess of collective pressure *and* a deficit of collective goals can lead to the loss of members and the collapse of a green community. Sociocracy solves this conundrum between community control and personal freedom by engaging in a dialogue that integrates everyone's concerns in order to transform and improve a given project until it becomes acceptable to everyone. Nothing is imposed by the community or by any individual without the consent of everyone. Do-ocracy solves the same conundrum by decentralising decision making to smaller groups, at a level where one no longer finds disagreement or minority views, and consensus flows naturally without even being discussed. So we are now back to the question of scale, since the segmented logic of do-ocracy works best at the lowest scales of social organisation.

Finally, the issue of private investment (in time and money) and the presence or absence of common resources is crucial in determining whether decision making is based on majority rule or takes on more consensual forms. When private investments or resources held in common increase, so does the push towards majority rule and formal decision making, from both external institutions and internal members. What is at stake is too important to risk being taken hostage by a few individuals.

In the absence of private investment or common resources, it is easy for members to solve a conflict by 'voting with their feet' and leaving the community if they are not satisfied with the way it is run. This is much more difficult when there are high levels of private investments or common resources, as leaving the community might signify losing one's investment or stake in the common resources. A formal and transparent democratic process based on majority rule secures that everyone's voice is heard and has an equal weight. Do-ocracy, for its part, works best when no common resource is involved. The lower the stake, the easier it is to consent to anything.

In the examples discussed above, living in an intentional community means sharing many resources and having to agree on collective rules. Since housing prices in many intentional communities are regulated and disconnected from the national housing market, residents might not be able to afford moving out of the eco-community to buy another home at full market price if they disagree with their neighbours on how to run the community. In this case, the stakes of collective decision making are extremely high, not just financially but also for the general satisfaction and well-being of residents.

Things are totally different in a green neighbourhood where members manage no (or negligible) material or financial resources collectively and can leave the group at any time without any consequence. In these circumstances, do-ocracy can be a convenient, efficient, and sufficient decision making mechanism. But as soon as substantial resources are shared collectively (and it can start with as little as bee equipment costing 45 euros per member), conflicts arise and more formal and inclusive forms of decision making such as sociocracy enter the picture.

Conclusion

The practice of a certain type of micropolitics has a deep impact on social dynamics. Green communities are constantly trying to maintain a balance between social control and personal freedom and a balance between efficiency and inclusion. In other words, how do these communities reconcile the frustration generated by a time-consuming decision making process and the frustration caused by not being heard or taken into account in the decision making process?

Green communities solve these conundrums by relying on hybrid micropolitics that combine different approaches. While the quest for consensus is a widespread ideal that helps maintain social cohesion that can be reached by diverse avenues (including sociocracy and do-ocracy), majority rule may remain a necessity when interacting with bureaucratic structures when the community grows beyond a certain scale or when significant resources are shared collectively.

Although the term 'democracy' is often associated with majority rule, it must be broadened to include other forms of micropolitics that facilitate collective decision making and avoid the concentration of power in the hands of a few. Sociocracy and do-ocracy are highly democratic forms of micropolitics serving that purpose while at the same time securing agility in decision making and social cohesion in the community.

References

Barnette, M., & Barrett, G. (producers). (2015, 20 February). 015. 'Do-ocracy'. A way with words [Audio Podcast]. https://soundcloud.com/waywordradio/1417-do-ocracy

Bertrand, G. (2022). *Governance and decision-making: Lessons from three French ecovillages* [Unpublished master's thesis]. KTH Royal Institute of Technology.

Blijleven, W. (2016). *Accountable do-ocracy: An exploratory study of the accountability practices of active citizens* [Master's thesis]. Utrecht University.

Buck, J., & Endenburg, G. (2010). *The creative forces of self-organization*. Sociocratic Center.

Buck, J. A., & Villines, S. (2017). *We the people: Consenting to a deeper democracy*. Sociocracy.info.

Chen, K. K. (2009). *Enabling creative chaos: The organization behind the Burning Man event*. University of Chicago Press.

Christian, D. L. (2014). Why no tyranny of the minority in sociocracy: How sociocracy can help communities. *Communities*, 165, 60–63.

Christian, D. L. (2016). Avoiding 'sociocracy wars': How communities learn sociocracy and use it effectively…or not. *Communities*, 173, 58–61.

Clemmensen, N. (2008). The development and structure of associations in Denmark, c. 1750–1880. *Scandinavian Journal of History*, 13, 355–370.

Comte, A. (1851). *Système de politique positive* [System of Positive Polity]. Carilian-Goeury & Dalmont.

Czekaj, M., Walczak, M., & Zięcbicki, B. (2020). Towards the sociocratic organization model. *Organization Review*, 10(969), 13–19.

Dietvorst, L. (2024). *Degrowth and eco-villages: What can social housing corporations learn?* [Unpublished master's thesis]. University of Groningen.

Endenburg, G. (1974). *Dictatuur, democratie, sociocratie* [Dictatorship, democracy, sociocracy]. Endenburg Elektrotechniek.

Endenburg, G. (1975). *Sociocratie: Een redelijk ideaal* [Sociocracy: A reasonable ideal]. Woudt.

Endenburg, G. (1998). *Sociocracy as social design: Rationale of a new social design for society*. Eburon.

Englyst-Nielsen, C., & Gausset, Q. (2024). From countercultural ecovillages to mainstream green neighbourhoods: A view on current trends in Denmark. *npj Climate Action*, 3, Article 60. https://doi.org/10.1038/s44168-024-00140-2

Gausset, Q. (2020). Stronger together: How Danish environmental communities influence behavioural and societal changes. In J. Hoff, Q. Gausset, & S. Lex (Eds.), *The role of non-state actors in the green transition* (pp. 52–70). Routledge.

Gausset, Q., & Hoff, J. (2024). Fællesskab – Et overset element i den grønne omstilling [Community – An overlooked element in the green transition]. In S. Krøijer, & L. Tønder (Eds.), *Grundbog om grøn samfundsteori* [*Textbook on green social theory*] (pp. 367–383). Gyldendal.

Grubb, A., & Henriksen, L. S. (2019). On the changing civic landscape in Denmark and its consequences for civic action. *Voluntas*, 30, 62–73.

Gundelach, P. (1984). Social transformation and new forms of voluntary associations. *Social Science Information*, 23, 1049–1080.

Gundelach, P., & Torpe, L. (1997). Social reflexivity, democracy and new types of citizen involvement in Denmark. In J. W. van Deth, & H. Keman (Eds.), *Private groups and public life: Social participation and political involvement in representative democracies* (pp. 47–63). Routledge.

Hansen, A. H. (2019). Et bofællesskab og en bevægelse: En antropologisk analyse af økosamfundsaktivisme på tværs af skalaer [A co-housing and a movement: An anthropological analysis of ecovillage activism across scales]. *Politik*, 22(3), 99–114.

Hansen, A. H. (2020). It has to be reasonable: Pragmatic ways of living sustainably in Danish ecocommunities. In J. Hoff, Q. Gausset, & S. Lex (Eds.), *The role of non-state actors in the green transition: Building a sustainable future*. Routledge.

Henriksen, L. S., & Bundesen, P. (2004). The moving frontier in Denmark: Voluntary–state relationships since 1850. *Journal of Social Policy*, 33, 605–625.

Hoff, J., Gausset, Q., & Lex, S. (2020). Introduction. In J. Hoff, Q. Gausset, & S. Lex (Eds.), *The role of non-state actors in the green transition: Building a sustainable future* (pp. 1–13). Routledge.

Højgaard, C. D. (2020). *Elastic integration in civic do-ocracies: How organizing shapes collective civic engagement in informal, non-membership-based networks*. In *Paper Presented at the 36th European Group for Organizational Studies (EGOS) Colloquium*, Hamburg, Germany.

Laloux, F. (2014). *Reinventing organizations*. Nelson Parker.

Owen, R. L., & Buck, J. A. (2020). Creating the conditions for reflective team practices: Examining sociocracy as a self-organizing governance model that promotes transformative learning. *Reflective Practice*, 21, 786–802.

Priest, J., Bockelbrink, B., & Liniana, D. (2024). *A practical guide for evolving agile and resilient organizations with Sociocracy 3.0*. https://sociocracy30.org/_res/practical-guide/S3-practical-guide.pdf

Rios, M. (2011). Sociocracy: A permaculture approach to community evolution. *Communities*, 153, 20–23. www.sociocracyforall.org/

Selle, P., Strømsnes, K., Svedberg, L., Ibsen, B., & Henriksen, L. S. (2019). The Scandinavian organizational landscape: Extensive and different. In L. S. Henriksen, K. Strømsnes, & L. Svedberg (Eds.), *Civic engagement in Scandinavia: Volunteering, informal help and giving in Denmark, Norway and Sweden* (pp. 33–66). Springer.

Simmonds, M. (2016). *Reconomy guide: The Do-Ocracy handbook*. Transition Network.

Straub, V. (2021). *Applying ecovillage praxis to foster regeneration and urban socio-ecological transformation: An exploratory analysis of five Malmö community gardens* [Unpublished master's thesis]. Lund University.

Tabor, H. (2024). *How sociocracy is applied in the Dutch ecovillage: Insights from Dutch ecovillage inhabitants and sociocracy experts* [Unpublished master's thesis]. University of Twente.

Tijhuis, K. (2021). *Creating an ecovillage: The trials and tribulations of building a sustainable community* [Unpublished master's thesis]. Utrecht University.

Torpe, L. (2003). Democracy and associations in Denmark: Changing relationships between individuals and associations? *Nonprofit and Voluntary Sector Quarterly*, 32, 329–343.

Vrielink, M. O., Verhoeven, I., & van de Wijdeven, T. (2013). Meedoen met de overheid? Over de stille beleidspraktijk van de doe-democratie [Participating in government? On the silent policy practice of do-ocracy]. *Bestuurskunde*, 22(2), 13–22.

Ward, L. F. (1883). *Dynamic sociology*. Appleton & Co.

Yang, J. (2021). *Perception of participation in decision-making: A case study of Ecovillage Boekel* [Unpublished master's thesis]. Masaryk University.

Zarinah. (2016, April 25). *How to nurture a living and evolving do-ocracy*. Embassy Network. https://medium.com/embassy-network/an-evolving-doocracy-3a6123f9b170

4

COMMUNES, ECO-VILLAGES, ECO-HABITATS, AND ALTERNATIVE NETWORKS

A typology of green communities and the role of subsistence in ecological living

Ieva Snikersproge

Introduction

In French alternative circles, it is widely assumed that any alternative to capitalism must be collective. Likewise, activist-scholars often share the belief that grassroots collectives are more knowledgeable, better organised, and more capable of proposing and implementing solutions to a wide variety of societal problems. However, the precise meaning of 'collective' is rarely problematised, theorised, or clearly understood. Not all collectives are equally collective, which affects their transformative potential.

This chapter seeks to expand our understanding of green communities specifically and various collective forms more broadly by examining their emancipatory potential and their capacity to contribute to ecological transition and transformations of ways of life. It draws on fieldwork conducted among 'back-to-the-landers', referred to locally as 'neorurals'.

For over 50 years, back-to-the-landers in France have been at the forefront of ecological innovation, pioneering organic farming, ecological housing, and the re-localisation of production in areas such as clothing and agricultural machinery. To implement these initiatives, they have experimented with a wide range of collective organisational models, offering valuable insights into the advantages and drawbacks of different collective forms.

The typology I present does not classify groups based on ideological distinctions, such as 'anarchist', 'feminist', or 'ecological community'; nor do I classify them by the governance structures they adopt (see Metcalf 1984; Meijering et al. 2007). Instead, the typology I use categorises ecosites and intentional communities into four groups: the commune, the eco-village, the eco-habitat, and individual households in alternative networks, according

DOI: 10.4324/9781003769064-5

to what they collectivise or share (see also Escribano, Lubbers, & Molina 2020). This classification illustrates that the more a community shares, the stronger its break from mainstream society and the greater its reduction in carbon footprint.

However, deeper collectivisation also results in more complex and demanding participation. While full-scale communes no longer hold mass appeal in France, they offer valuable lessons for understanding modern society's entrenchment in an ecologically unsustainable dependency on growth. Difficult access to property, combined with the devaluation of subsistence-related labour, creates systemic obstacles to the development of eco-villages and other green communities, while locking people into more individualised and carbon-intensive livelihood strategies. Most importantly, this typology demonstrates that collectives providing access to essential means of subsistence are more robust than those formed around less critical aspects of life and have a stronger ecological impact on ways of living.

To shed light on the systemic pressures toward decollectivisation in France and other modern capitalist countries and the way these pressures restrict attempts to develop more ecological forms of living, the second part of this chapter examines the case of Peacock, an ecosite in southeastern France that evolved from a commune to an eco-village and later to a hybrid eco-habitat with a guesthouse as a source of income. Finally, I conclude with some policy recommendations to foster the formation of green communities and promote a transition to ecologically sustainable ways of life, while balancing individual freedom with security within larger collective structures such as markets, welfare programmes, and labour laws.

Green communities and the French back-to-the-land movement

In 2014, when I first arrived in Diois, a vast rural area in the Prealps, I spoke with as many people as possible and expressed my interest in studying eco-villages as potential laboratories for ecological living in a (future) post-carbon society. I also told them that I hoped that as my interlocutors they would help me identify ecological initiatives in the area that were worth studying.

One of my first contacts was Jean, a co-founder of Peacock, an eco-place I describe in more detail later in the chapter. His response to my inquiry was a wide grin of disbelief: 'I do not know of a single eco-village in Diois! The closest thing to a commune I've seen is a collective in the neighbouring village, and it's more like a shared flat'. I met Jean while visiting with a group interested in creating an eco-village; he recommended we read the comic book *La Communauté* (*The Commune*) by Yann Benoît and Hervé Tanquerelle (2010), which recounts the trajectory and eventual dissolution of a commune formed in the Loire Atlantique department after the nation-wide protests in May 1968.

Jean's comment, shaped by his own experience at Peacock, which had gradually scaled back its collective ambitions, reveals that while both an eco-village and a shared flat are 'collective', there is a significant difference between them. The typology I present here—the commune, the eco-village, the eco-habitat, and individual households within alternative networks—reflects terms and distinctions I encountered during my fieldwork among back-to-the-landers and other ecological alternatives in France.

Since 2015, I have conducted intermittent fieldwork in Diois, a historic area of neorural settlement. Over the years, I have participated in numerous workshops and festivals in the area, conducted around 140 semi-structured interviews, and followed dozens of collective initiatives and several collective ecological living arrangements.[1]

While this typology resonates with the terminology used by individuals engaged in ecological alternatives, the terms were rarely employed in a consistent or clearly defined manner in France. On the contrary, terms like 'écolieu' (ecosite) and 'oasis' were often used to group together a wide variety of collectives aiming to live in environmentally conscious ways.

The origin of the term 'ecosite' is hard to trace—it refers directly to a place centred on ecology. But it has been popularised by *Passerelle Éco* (which roughly translates as 'Eco Gateway' or 'Eco Bridge'), an independently published, anonymously edited magazine launched in 1999. The magazine promotes radically ecological lifestyles and connects people interested in creating eco-villages.

The term 'oasis' was introduced by Pierre Rabhi, a French peasant, essayist, and agro-ecology activist of Algerian origin, in his 1997 manifesto *Oasis en tous lieux* (*Oasis in Any Place*). Initially, the manifesto garnered little attention; after its publication, only about ten communities self-identified as oases and at least partially met the manifesto's demanding criteria for autonomy. However, since the 2010s, there has been a notable proliferation of oases, especially after 2018, when Colibris, a movement also founded by Rabhi that gained more traction in urban areas, began actively promoting the oasis concept by creating a cooperative. This cooperative supports new projects, disseminates how-to information, and assists with finding funding for the creation of new oases.

The cooperative also broadened the definition of an oasis by establishing five core principles: sharing and pooling of resources, governance that respects individuals, openness to the outside world, sobriety, and food sovereignty. As of 2025, the cooperative's website lists over 1,200 oases that encompass not only eco-villages but also ecological cohousing projects, green coworking spaces, and community recycling initiatives. While oases continue to proliferate, not all eco-villages appear in the national oasis database. When I asked the inhabitants of these places why they had chosen to remain outside the nation-wide movement, they described this as a deliberate decision. They either wanted to distance themselves from Rabhi or they perceived the oasis movement as too closely aligned with 'cheerful' green capitalism that promotes some ecological gestures such as recycling and solar panels but does not question the need for economic growth and its toll on the planet. They wanted to step away from what they saw as an ideal of joyful, poorly politicised ecological living that often costs considerably more than conventional housing.

Both ecosite and oasis resemble the English term 'intentional community', which Geoph Kozeny defines as 'a group of people who have chosen to live together with a common purpose, working cooperatively to create a lifestyle that reflects their shared core values' (Kozeny 1995: 1; see also Sargent 1994: 15; Brown 2002: 3; Sargisson 2012: 31). These broad terms emphasise the voluntary, purpose-driven nature of communities built around alternative ways of living. However, they do not distinguish between different types of intentional communities, and many differences do exist. Many researchers of intentional communities have observed their gradual individualisation and deradicalisation, followed by recent re-radicalisation (Meijering 2012; Sargisson 2012; Rouvière 2015; Magnusson 2018; Halfacree 2022). In other words, many such communities have shifted from relatively self-sufficient collective communes to forms of ecological cohousing integrated into the market economy, while more recently there has been a renewed push for more radical, degrowth-oriented projects. I argue that a typology of ecosites grounded in what is shared and what remains individual offers insight into this historical trend of green communities, its implications for ecological transition, and the centrality of subsistence in radical projects.

The four types of collective projects

The commune

A commune follows the organising principle articulated by Louis Blanc (1839): 'From each according to his faculties, to each according to his needs'. In practice, this creates a family-like organisation that does not strive for reciprocity or exchange with expected equivalence. Communes are often described as 'communities of work and life' that blur the line between work and leisure in order to avoid creating hierarchies among different types of labour such as caregiving and household maintenance, whether the work is agricultural, political, or reproductive.

Communes tend to remain loosely organised. Everything—living space, equipment, and money—is shared, often without clearly defined boundaries regarding responsibilities. However, as in families, implicit norms and limits usually develop in order to enable a workable equilibrium. As one of my interviewees explained, these kinds of communes require socially skilled individuals who know how to balance their contributions to the collective with their own needs. They must be able to decline unreasonable demands clearly and constructively. In practice, interactions among residents in such communities are shaped not only by explicit verbal requests but also by lived, evolving actions that shape the rules of conduct within the collective as a whole.

Because communes develop organically in response to emerging everyday situations, they tend to generate a steady stream of minor but manageable conflicts. For example, one of my interlocutors told me about a disagreement over a member buying a T-shirt for 5 euros with communal funds. The member justified the purchase by saying it was inexpensive and necessary, while others objected that such a low price implied exploitative labour conditions abroad, something the group fundamentally opposed and did not want to endorse with its money. This example illustrates how French alternative circles often 'politicize the smallest gesture' (cf. Pruvost 2015), which can create conflicts that are usually addressed through meetings and collective discussions.

However, some disagreements lie beyond the reach of dialogue and consensus and potentially lead to the dissolution of the collective. Conflicts often arise from the partial and situated nature of social knowledge because the social world is never fully available to us as a total object of study, which gives rise to divergent interpretations and worldviews.

A common point of rupture is differing views on how radical the collective should be, i.e., whether it should pursue full autonomy from the market economy and the state that some see as inherently complicit with capitalism. Since 'autarky', i.e., full self-sufficiency, is nearly impossible to achieve, all collectives more or less implicitly accept compromises. A common way to gloss over differing red lines in terms of engagement with the market and the state is to play down contentious issues during a group's formation by focusing on practical, consensus-building projects and avoiding sensitive ideological topics. Focusing solely on practical issues is only a weak solution, however, since controversial questions tend to reappear over time. They must either be persistently sidestepped or, if openly debated, can result in either a workable consensus or the dissolution of the collective.

For example, during the COVID-19 lockdowns, I observed sharp divisions over whether to comply with state-imposed distancing measures. Some members deeply distrusted state institutions and mainstream media, and others expressed greater concern about the virus itself. In the collective I was following at the time, the group temporarily split into two subgroups: one group strongly opposed what they perceived as a dictatorship of the state and pharmaceutical companies and completely disregarded social-distancing measures. Another group adopted strict self-isolation out of concern for the elderly and the risk of overloading public hospitals.

Although the collective endured beyond the pandemic, these divisions left lasting traces. Some members came to be seen as overly complicit with the state, others as prone to conspiracy theories, which made the issue into a sensitive topic, which must largely be avoided in order to continue acting together.

Pure, ideal-type communes are rare today. Most first-wave communes from the 1960s and 1970s disbanded; scholars estimate that roughly 95% of these communities have dissolved, while the remaining 5% have evolved into more structured organisations (Hervieu-Léger & Hervieu 2005). This high failure rate is largely attributed to the fact that ostensibly horizontal decision making structures can conceal informal hierarchies that are often led by charismatic male figures and that the embrace of sexual liberty in a community can sometimes create emotional entanglements that are incompatible with rational, collective governance. Today's collectives often strive to distance themselves from the widespread stereotype of these early communes, which have been ridiculed in the popular imagination.

Still, commune-like structures persist among close-knit friend groups or in temporary collaborations around specific tasks, such as organising festivals or undertaking large construction projects. These two types of cases differ significantly. Some friend groups function like communes because of their homogeneity: they share the same worldview and similar desires, lifestyles, and daily rhythms. Also, commune-like experiments tend to flourish in squatter communities and already established green communities that do not need to worry about paying a mortgage. However, none of these similar qualities protect them from potential future internal conflict. Sharing requires a continuous emotional and relational labour (see Penny Clark's chapter 7 for a more detailed discussion of this issue).

By contrast, adopting commune-like practices for short-term, goal-oriented activities is much less demanding than full-time participation in a 'community of work and life'. These temporary arrangements are stronger examples of what Graeber (2010) calls 'everyday communism': the spontaneous, non-reciprocal sharing of labour and resources to achieve common goals. Yet because everyday communism is limited in scope and does not reorganise the entire lives of its temporary participants, its transformative potential remains modest.

Ideal-type communes that are both collective and highly autonomous tend to excel ecologically. They dramatically re-localise the production of essential goods and services, and what they do not produce themselves is often acquired second-hand. Choosing to use second-hand goods is ecologically advantageous for two reasons: (a) refurbishing existing items is less environmentally harmful than buying new ones, and (b) such items often involve less resource-intensive technologies that have been displaced by more modern, labour-efficient alternatives (e.g., manual juice presses and simple mechanical tractors). Second-hand goods not only align with the ecological convictions of these communes but, also, communes have limited monetary resources.

However, a high level of autonomy comes with trade-offs, including limited access to advanced technologies and fewer economies of scale. As a result, these communes exhibit a smaller social metabolism per inhabitant, i.e., the energy and resources required to sustain the life of a single person when compared to the broader society. The downside is that communes are labour-intensive and provide a lower standard of living if judged by residents' disposable income. Nevertheless, many commune members, like those of eco-villages and eco-habitats, report an improved quality of life, more meaningful work, a stronger connection with nature, a sense of community, and the ability to live according to their ideals. (For similar findings on green communities in the Danish context, see Gausset & Jensen 2024).

The eco-village

The second type of collective project, eco-villages, could be understood as a more structured version of communes. They collectivise aspects of work and living as communes do, but there are clearer rules and limits that entail some degree of 'deradicalisation' by shifting larger parts of everyday life into the private sphere. For example, if eco-villagers have fulfilled their duties vis-à-vis the eco-village, they can use their time as they see fit. Eco-villages prioritise shared economic activities, which may include some reproductive tasks such as shared meals and child-rearing while maintaining relatively conventional private lives that are often organised around nuclear families. As a result, large segments of care work are carried out by families and, despite aspirations to gender equality, often by women.

Clearer entry and exit rules that often involve the buying and selling of shares in the eco-village also allow people to leave a project more easily if living there no longer aligns with their ideals. The ability to exit without losing all invested time, energy, and resources is generally welcomed because staying together out of necessity rarely fosters a thriving social organisation. In reality, however, selling shares in an eco-village is never simple: prospective buyers must align with the worldview and rhythm of the eco-village. While it is theoretically easier to join an eco-village than a commune due to its less comprehensive character, eco-villages nonetheless rely on shared ideas about the degree of rupture, or differentiation, they seek to establish vis-à-vis mainstream society. Moreover, disagreements often arise over the fair price of eco-village shares. On the one hand, departing members may wish to sell their shares close to market value in order to secure housing elsewhere.

On the other hand, remaining members prefer to keep shares affordable and resist the rise of property prices that can fuel speculation and contradict their values.

Economic activities in eco-villages typically fall into two categories: income-generating work for external necessities (such as mortgage payments, but also cars and computers) and subsistence work aimed at greater autonomy (see also Sylvia Amar, chapter 8). Some eco-villages earn income by selling agricultural products or services as early communes once did. But increasingly, they rent out premises for yoga and wellness retreats or offer workshops on permaculture and collective governance because these activities tend to be more lucrative than farming.

Balancing money-making and subsistence work can be difficult: it is often more efficient to earn income through educational workshops than to sustain oneself solely through permaculture-based farming (Snikersproge 2022). As a result, subsistence activities may be neglected. They can also be strained by the increased food demand that is created by visiting workshop participants, volunteers, and others. When eco-villages charge high fees for permaculture or farming workshops but rely on purchased food themselves, some outsiders perceive this as hypocritical.

From an ecological perspective, eco-villages aim high: income-generating work is ideally in service of a relatively autonomous, subsistence-oriented way of life. Most eco-villages meet their food, energy, and housing needs locally, which significantly reduces their social metabolism per inhabitant when compared to mainstream society. However, greater emphasis on income generation—and the possibility for individuals to maintain separate economic activities—generally results in a somewhat higher social metabolism than that of communes, which rarely carry mortgage debt and impose stricter limits on individual economic accumulation.

Most 'how-to' literature on collective living focuses on models like eco-villages (and the eco-habitat discussed below) and offer guidance on creating deliberative spaces, articulating shared visions, structuring decision making systems, and managing conflict (see, for instance, Leafe Christian 2006; McCurdy 2013; and Gausset & Nielsen-Englyst chapter 3 and Schwab, chapter 6 in this book).

Yet eco-villages remain rare because of several challenges. Firstly, they demand significant time and labour, particularly for subsistence activities, which require a commitment that goes beyond conventional work hours. Second, property access poses a barrier, as eco-villages often face difficulties repaying loans. (For challenges inherent to class differences, see Esteves, chapter 2 in this book). Finally, the planning and construction phases, from defining goals and governance to renovating or building dwellings, typically span a decade or more.

The eco-habitat

The third type of collective project is the eco-habitat. Like eco-villages, eco-habitat projects have clearly defined rules that distinguish the individual from the collective and outline their vision, goals, rights and duties, and procedures for entry and exit. However, unlike eco-villages, eco-habitats do not engage in shared economic activities: they leave these decisions to individual discretion. As a result, most inhabitants lead fairly conventional economic lives: they run businesses, work salaried jobs, or receive pensions or welfare benefits. Some engage in subsistence gardening, but it is typically a form of activism or a hobby rather than a livelihood.

In terms of housing, eco-habitats resemble eco-villages: they combine individual homes with shared spaces and follow collectively agreed-upon principles such as ecological construction and environmentally friendly practices like waste sorting and car sharing. While shared spaces may require occasional collective maintenance, participation in communal projects is generally optional, given the absence of shared economic activities. Those with more time, such as retirees or parents on leave, tend to be more involved than those balancing demanding professional and family lives. Although physical proximity facilitates mutual help, domestic life largely remains centred on family units.

Because eco-habitat projects place relatively light demands on residents, they sometimes fall short of ecological ideals, which can spark tensions in the community. For instance, interviewees expressed frustration when neighbours ordered from Amazon or bought industrially produced ham from the supermarket instead of supporting local farmers. While these consumption choices were often attributed to time or financial constraints, there was still an underlying expectation that residents would make more of an effort towards adopting ecological behaviour.

For example, some residents suggested that reducing meat consumption could make local, organic food more affordable. In response to such unspoken tensions, many shrugged and said, 'We are not a sect', emphasising individual freedom over uniformity in lifestyle.

Overall, eco-habitat members lead more ecological lives than the average French citizen. They live in eco-friendly homes powered mostly by renewable energy and share certain spaces, tools, and vehicles to reduce each household's ecological footprint. Many eco-habitats also follow a guiding charter and cultivate a normative, though sometimes contested, culture that encourages sustainable consumption. However, since each household remains economically autonomous, there is less coordination in relocalising subsistence and a higher reliance on market-based—and sometimes less ecological—solutions for everyday needs. Conventional lifestyles may also increase work-related consumption such as clothing, technology, and commuting and contribute to time poverty, which causes a reliance on less ecological conveniences like ready-made meals and dry cleaning. In general, eco-habitat residents tend to have less ecological lifestyles than those living in eco-villages and communes.

Eco-habitat projects are becoming more popular. However, they are not part of the social housing sector and demand significant financial and labour investments due to the higher costs of ecological construction, which makes them less accessible. Participants are aware of this limitation and strive to demonstrate that eco-habitats need not be exclusive to the wealthy. Some aim to reduce costs by participating in construction themselves. Additionally, collective eco-habitat projects can offer real economies of scale. Cohousing enables groups to acquire properties that are too expensive for individual households, bypass developers, and access land at lower costs. These projects can also negotiate bulk discounts on materials and share labour during construction. Some, like certain eco-villages, also attempt to limit capital gains when reselling shares in the project. Because eco-habitats are less all-encompassing than eco-villages, property turnover is more common, which makes real estate speculation a more pressing concern.

Since eco-habitat projects primarily share living spaces and rely on voluntary collective involvement, some outsiders criticise them as 'bourgeois' rather than genuinely alternative. One interviewee dismissed a neighbouring eco-habitat, saying, 'There are collective projects where the only unifying factor is that everyone sorts their trash and does yoga on Sundays'. By calling the project bourgeois, the interviewee not only pointed to its exclusivity but also to its weak sense of collectivity: participation in communal activities depends on whether individuals have the time, money, and energy to contribute.

Individual households in alternative networks

The fourth type of collective project could be described as individual households in alternative networks. These are collective organisations that allow participants to pool resources to access goods and services that might otherwise be inaccessible due to lack of money, ineligibility for state services, or because such usually more ecological or otherwise 'ethical' goods and services simply are not produced elsewhere. Examples include time banks, pay-what-you-can systems, community bars, work parties, construction associations, free shops, cooperatives, carpooling initiatives, collectively managed gardens, community-supported agriculture, and collective protests.

Alternative networks are the most widespread form of collective project because they are relatively easy to join. They do not require the full-time, holistic commitment expected in communes or eco-villages, and to a lesser extent, eco-habitats. While some require regular meetings or voluntary labour, entry and exit are usually easy because most do not involve property ownership. When property is involved, it is often lent by local authorities or benefactors, which removes the need for significant personal financial investment.

As these networks institutionalise and professionalise, they often become associations that seek state funding (for a detailed analysis of an alternative food network, see Cristina Grasseni, chapter 10 in this book) or cooperatives that, while socially oriented and imposing limits on capital gains, function much like conventional enterprises with fixed capital and salaried employees. The more institutionalised the alternative, the more it develops its own rules about what remains and what can be individually appropriated, which ultimately resemble more conventional models such as associations and cooperatives.

Non-professional alternative networks often struggle to attract enough active participants to drive meaningful change. My research shows that while these networks support the livelihoods of many Diois residents, they rarely constitute their main source of livelihood. As with eco-habitats, people's income shapes their availability for participation in alternative networks and labour-intensive ecological activism. Income also affects consumption choices: lower-income households might opt for cheaper, industrial goods with higher environmental costs, while wealthier households can afford more ethical and ecological products but tend to consume more overall, which counterbalances the benefits of 'ethical' consumption (Dubuisson-Quellier 2010; Coulangeon et al. 2023). Although these networks aim to bridge such gaps, they tend to attract people of similar socio-economic backgrounds with shared concerns, ideologies, income levels, and daily rhythms (Snikersproge 2023).

Many alternative networks in Diois offer effective ecological solutions by recirculating materials, promoting low-impact housing, reducing car use, encouraging repair, and democratising access to organic, small-scale peasant-produced food. Yet, they remain compartmentalised: in theory, they are open to all, but in practice, participation requires a particular mix of strong ideological commitment, financial constraints, and available time.

Households with a deep commitment to ecological sustainability and a high level of subsistence activities often participate actively in these networks and reduce their ecological footprint to levels comparable with eco-villagers or commune inhabitants. Geneviève Pruvost (2021, 2024) has described how focus on subsistence—through self-production of housing, energy, and food (vegetables, eggs, meat, cheese) and the re-use of clothing, tools, and materials—can serve as the foundation for an emancipatory, ecological lifestyle. She also demonstrates that in present-day France, such ways of living are not limited to the wealthy.

My research confirms that many neorural households adopt subsistence-centred lifestyles at the household level to avoid conflicts that are common in communes and eco-villages. However, this also creates tensions within couples. While it may seem easier to manage the reinvention of a way of life in small social units like couples, where uneven task division can be more readily accepted for the sake of household efficiency, they can still collapse under the burden of these tasks. At the same time, greater autonomy from the market economy, such as having secure housing, enables people to reclaim time, restructure their lives, and engage more fully in alternative networks to further reduce their ecological impact.

Comparison

When we consider these four types of collective projects, we observe a general trend toward individualisation. While this makes participation easier, it also limits the transformative potential of these projects. The commune integrates all aspects of life; the eco-village shares economic activities and living space while setting limits to collective expectations; the eco-habitat focuses on shared living spaces and individualises economic activity; and alternative networks offer punctual or compartmentalised collaboration with clear boundaries in different, precisely delineated spheres of life and/or work.

What becomes increasingly de-collectivised in all types of collective living is access to the means of subsistence, which means that individuals are left on their own to find the resources necessary to participate in a collective. De-collectivisation makes these collectives less demanding in terms of work rhythm, prioritised activities, and the standard of living, which facilitates participation by more diverse types of people. But it also makes these collectives either optional or subordinated to the time and resources that people manage to liberate from their jobs and family life.

Table 4.1 compares the four types of collectives in terms of four qualities, from organisation of daily living to the degree of separation of a given type of collective project from conventional society.

TABLE 4.1 A typology of green communities: Ease of participation and the potential for ecological transformation

	Things placed in common	*Ease of participation*	*Frugality*	*Rupture from mainstream society*
Commune	+ + +	- -	+ + +	+ + +
Eco-village	+ +	-	+ +	+ +
Eco-habitat	+	+	+ / -	+ / -
Individual households in alternative networks	-	+ +	+ / -	+ / -

Of course, the four types are ideal types; in reality, collective projects mix different modalities that structure their functioning. For example, even the most structured eco-villages recognise that it is impossible and suffocating to have a clear rule for everything. Therefore, there are significant spheres of activity that continue to function like communes, i.e., without a strict distinction of rights and duties and a relative disinterest in balanced, symmetrical quid pro quo reciprocity.

Also, eco-habitat projects, commonly critiqued as bourgeois (not least by the participants themselves), want to go beyond simple co-living in luxurious, ecological housing. Many open their eco-habitat to outsiders for various collective projects that they see as worth supporting. For instance, they lend their collective areas to a handicapped people's association or host volunteers who help with organising a festival on ecology. They might also help other co-members of the eco-habitat when they run into economic and other types of difficulty or join in collective efforts to organise a festival that has both ideological and financial aspects.

However, although there are no clear-cut lines between communes, eco-villages, eco-habitats, and other collective alternatives, they do reflect one general tendency: while the most widespread collective projects do not collectively manage the means of subsistence, they tend to individualise and de-radicalise.

Whether a collective project manages subsistence activities as a commons-based institution or not marks a significant qualitative difference between different intentional and green communities. Communes and eco-villages, two ideal-type green communities that seek to increase the autonomy of its inhabitants by focusing on subsistence activities, provide the biggest rupture with conventional society and also the greatest reduction in ecological footprint. This qualitative break between communes and eco-villages on the one hand and eco-habitat and alternative networks on the other is confirmed by other researchers.

For instance, Lucy Sargison (2012) describes cohousing (eco-habitat) projects as 'second generation' intentional communities that are less ideological than 'first generation' intentional communities, i.e., communes. Madaleine Sallustio (2022) defines neo-rurals or back-to-the-landers not as all urban-to-rural migrants, which, to her, empties the term of its sense, but as people who take up agriculture (a type of subsistence activity) in a collective setting. However, while such 'radical' green communities persist, they remain few, which is not only due to the lack of ideological commitment, but also to omnipresent structural obstacles (see also Nielsen-Englyst & Gausset 2024).

Case Study: Peacock

To better illustrate the challenges of constructing radical collective projects, I present the case of Peacock, an eco-place that transformed over time from a commune-like organisation to an alternative co-living or eco-housing arrangement.

Peacock was founded by a group of six friends who travelled together through Southern Europe and North Africa before settling down to draw inspiration from other eco-places and strengthen their communal ties. Jean, one of Peacock's senior members, recounted how their travels in North Africa tested their capacity for communal living as they shared the responsibility for setting up campsites, cooking, and cleaning. After their travels, they purchased Peacock, an old stone compound in northeastern France at 1000 metres above sea level that offers stunning mountain views but harsh, snowy winters.

Initially, the group pooled 120,000 euros for the down payment on the 600,000 euro property, with three members contributing 40,000 euros each. They opted for a collective property regime, which allowed each person to hold transferable shares. The bank approved their loan because the property was in good condition and because Peacock had a shared economic project: operating a guesthouse with room for yoga and wellness retreats. During the first year, their organisation was relatively fluid and resembled a commune. Jean described this phase as a 'honeymoon period' marked by collective euphoria.

It was after that first year that Steven, a Peacock member who worked hard to produce vegetables, began to feel that he contributed more than others. Jean recalled:

> Of course he worked a lot, but while working outside, did he see what others were doing? That there were people doing accounts or repairing the building? Cutting wood, or even playing the guitar to create quality moments together.
>
> *(Interview with Peacock resident Jean, February 2023)*

Above all, Jean believed that maintaining the building, renting it out, and keeping the accounts in order were more important than being self-sufficient in vegetables. Jean rhetorically asked, 'Would carrots help to pay down the debt? No, they would not!'

For Steven, it was unacceptable that food self-sufficiency should be subordinated to money-making, and he decided to leave Peacock with his partner, Anya. This first departure highlighted the issue that not all work was equally valued: the type of work Steven valued ideologically had the least macro-economic value. For Jean, since the loan had to be repaid, ideology came second to the financial viability of Peacock. Rather than asking for a reorganisation of the distribution of labour, Steven chose to leave. Jean added, 'We should also be realistic about how much work self-sufficiency entails. I don't want to spend my days like in Longo-Mai, working with my head between the handlebars all the time'. Longo-Mai, in Jean's opinion, requires too much field labour.

Longo-Mai is perhaps the most well-known and robust eco-village network in French alternative circles. Founded in the early 1970s by Swiss and Austrian anarchists inspired by the countercultural movement, its first and most renowned farm was established in Limans, France. Today, the network includes nine farms across France, Switzerland, Austria, Germany, and Ukraine. Initially driven by anti-authoritarian and anti-state ideals, Longo-Mai is now centred around two main pillars: (1) relative autonomy through subsistence work shared across its farms, including farming, preserving produce, repairing machinery, producing seeds, and converting wool into garments, and (2) political activism at national and European levels, particularly in support of agroecology and against far-right politics. Longo-Mai funds its political efforts by selling produce at Christmas markets, renting out guesthouses, and receiving regular contributions from a broad network of supporters, who in return receive a quarterly newsletter. This network allows Longo-Mai to fund both activism and new ventures that require major investments, such as semi-industrial equipment to produce wool clothing.

Steven and Anya's departure also highlighted the steep financial burden of collectively acquiring property. Few aspiring eco-villagers have enough inherited or accumulated wealth to purchase a large estate without incurring debt. In Peacock's case, the couple's departure meant losing the 80,000 euros they had contributed, which forced Jean to take out another loan to cover the gap. He proposed that other Peacock inhabitants acquire portions of property in exchange for work that would make the place run, but they preferred to live in and contribute to Peacock without becoming legal property owners. Many Peacock members were afraid that such an arrangement would be too constraining. They wanted to collectively manage the benefits that the place provided: shelter, workspace, heating with wood, garden access, occasional income, and proximity to like-minded peers, without assuming financial obligations. Ownership was seen as too demanding in terms of required income and labour.

Nonetheless, Jean decided not to abandon the project and took out the loan. He then became the sole person who owned portions of property in Peacock. After he took full ownership, residents discussed reorganising the community. Although collective expectations had been lowered, Peacock still required everyone's help. An informal agreement was made: each resident would pay a symbolic rent of about 30 euros/month and contribute one workday per week to maintain Peacock or run the guesthouse. According to Jean, this arrangement never really worked:

> When it was cold, nobody stepped up. When we organised a workday—say, to chop firewood—someone had a performance, someone else had to pick herbs blooming that week, and another had a sick mother. There were always good reasons for absence, but it was hard to coordinate work so that everybody would feel like the arrangement was just.
>
> *(Interview with Peacock resident Jean, February 2023)*

Moreover, Jean explained that when there were clients in the guesthouse, he had to be able to count on people who would cook, clean, change sheets, and do other errands to serve the guests. Because the proceeds from the guesthouse were used to pay back the loan taken out in his name alone, Jean proposed to pay wages for the hours that Peacock's inhabitants spent on the guesthouse. Thus, the functioning of Peacock resembled an eco-village less and less and an enterprise run by Jean more and more.

Of course, that was not Jean's intention: he wanted to keep the spirit of the original Peacock alive. The rent that Peacock's members were paying remained symbolic: they organised regular meetings to collectively discuss the functioning of the place and occasionally collaborated on collective projects such as cutting wood in the forest or organising an annual Peacock festival.

By recounting the trajectory of Peacock, I do not mean to imply that Peacock has become a non-alternative place that is not much different from the rest of the society or that this was the only possible outcome. On the contrary, ask anyone in Diois and they would still describe Peacock as an eco-place—a collective, alternative site. With its low rent and shared infrastructure, Peacock serves as a stepping stone for many neorurals who later go on to build other projects more closely aligned with their ideals. It is still seen as a valuable resource, and its residents continue to engage in alternatives beyond its boundaries. Its path is not fixed; the community evolves as people come and go. However, as Jean admitted, acquiring a share in Peacock is now much harder because of the increased property value.

Originally, Peacock aimed to collectively manage a living space that offered (1) shelter, (2) proximity to like-minded people to share life and projects with, (3) heating with wood, (4) food, (5) renewable energy, and (6) an enterprise to generate income for expenses like loan repayments. Today, it offers only (1) shelter, (2) proximity to like-minded people, (3) heating with wood, and (4) renewable energy. Peacock's inhabitants, whose number fluctuates somewhere between six and 15, have an environmental footprint significantly lower than that of the average French citizen but higher than it would have been had they continued producing food and enjoying a greater level of autonomy.

The ideal of food autonomy was abandoned. Residents contributed to purchasing staples like rice, sugar, and oil for the communal kitchen while buying comfort food individually. Guesthouse work was compensated with wages. In short, both the ownership model and the scope of what was shared changed. As economic activities—whether for subsistence or for the market—became individualised, Peacock's collective spirit weakened and collaboration became more ad hoc: people participated when they could and when they wanted to.

Peacock's story highlights two structural challenges to building green communities: access to property and the valuation of different types of labour. Property is expensive and hard to pay off with a subsistence-oriented way of life, which pushes people toward market-based work. This creates tension between income generation and autonomy-enhancing subsistence. It is often perceived that subsistence efforts come at the expense of care work or life-sustaining practices like eating, resting, caring, and spending quality time together. To determine sustainable levels of care work—whether it is paid labour or subsistence that undermines it—we must study communities that have scaled up and routinised subsistence effectively. What is clear, however, is that capitalism's valuation of labour fuels internal conflicts in collective projects that originally intended to propose alternatives, which contributes significantly to their decollectivisation.

What's collective in a collective project?

According to what I have observed in the French alternative circles, collectives or green communities differ in what they place in common: the more they share, the greater their capacity for transformative change toward ecological modes of life. Deradicalisation—understood as a diminished capacity to drive transformative change—results from the decollectivisation of these alternatives, particularly the decollectivisation of access to subsistence. Why? Because decollectivising access to the means of subsistence leaves participation in these alternatives up to individuals; it becomes more of a lifestyle choice, one that is subordinate to professional obligations. This increases work-related consumption and time poverty, which, in turn, elevates the social metabolism and carbon footprint per person. Focusing on subsistence is important because (a) when collective activity significantly contributes to livelihoods, people are more likely to commit, and (b) focusing on subsistence enables individuals to reduce their ecological footprint by relocalising production and minimising needs and work-related expenditures. However, it is difficult to prioritise subsistence because (a) it requires access to property, often at prohibitive prices; (b) this cost is impossible to offset through subsistence-related activity, which capitalism has devalued; and (c) balancing income-generating work with subsistence activities creates conflicts around the value of different types of labour—whether subsistence, paid, political, care, or recreational. This conflict raises issues of social justice concerning fair distribution of different types of work and leisure.

For collective, subsistence-focused projects to work, participants must have access to property, share a common analysis of the social world, and agree on a desirable standard of living and rhythm of everyday life. Since these conditions are difficult to align, most people choose not to share access to the means of subsistence, keeping it a private matter. This hinders participation in collective projects due to incompatible rhythms, priorities, and constraints, and prevents significant steps toward relocalising production, reducing needs, lowering ecological footprints, and collectively shaping desirable ways of living. As a result, neglecting subsistence as a collective issue leads to decollectivisation and the deradicalisation of intentional communities, which ultimately tether residents to energy-intensive lifestyles and the capitalist growth economy.

The case study of Peacock shows that the desire for individual freedom in organising daily life and defining acceptable living standards reinforces decollectivisation. However, this tendency is systematically encouraged by the capitalist market economy, which overvalues property, devalues subsistence labour, creates time poverty, and promotes work-related consumption. In other words, there is a systemic lock-in to the capitalist growth model.

The typology of four different types of collectives or green communities reveals both the importance of increasing autonomy by reclaiming access to the means of subsistence and the obstacles to doing so. While full-fledged communes tend to have limited appeal due to their constraints, homogenisation, and sometimes lower standards of living, the data presented here suggest that there are other ways to encourage more ecological lifestyles. Facilitating access to property and housing, re-valuing certain forms of human labour, and improving access to means of subsistence provide a promising starting point.

For example, it is possible to expand the social housing offer to include eco-habitats, which would increase their accessibility and almost by default reduce the ecological footprints of their inhabitants (see Kingfisher 2021). Another useful tool could be the creation of a special property regime for eco-villages, which would ease entry and exit across different projects while avoiding the overvaluation of property. In this context, one can draw inspiration from land community trusts or, more ambitiously, from Terre de Liens (cf. Lombard & Baysse-Lainé 2019), which combines an association, a foundation, and a real-estate company to establish a system of collective ownership for farms and farmland, thereby removing them from the mechanisms of capitalist market valorisation.

On another front, there are initiatives that contribute to the revaluation of human labour. For instance, well-designed cash transfer schemes such as an Ecological Basic Income (cf. Langridge, Buchs, & Howard 2023) could reduce people's dependence on the job market, thereby lowering work-related consumption, increasing time available for civic participation, and revaluing labour-intensive work such as care work or ecological labour (cf. Snikersproge 2024). Another example is social security for food (Chiffoleau et al. 2023), which, when well-designed, can revalorise the agricultural labour involved in producing agroecological goods while also improving economic security and freeing up time for civic engagement.

All the cited examples, apart from Ecological Basic Income, are existing experiments that are carried out largely by civil society, but their broader impact depends on stronger involvement from the state and/or municipalities. Access to property and the revaluation of different types of labour, particularly subsistence-related labour (see also Grasseni's chapter 10 in this book), are central to creating lasting green communities that invent ecologically sustainable modes of life.

Note

1 I conducted my fieldwork in the Diois region in stages. Between 2016 and 2018, I observed everyday life in the area through participation in public meetings, debates, work parties, organic farms, eco-villages, festivals, and cooperatives, and conducted interviews with around 80 '*néoruraux*' (neorurals) active in these alternative networks. From 2020 to 2023, I sought to situate these alternatives within the broader territorial economy by engaging in participant observation at municipal council meetings while carrying out a representative quantitative survey of 408 household responses covering 934 of the region's 12,000 inhabitants. I also conducted follow-up interviews: 30 with elected officials and 20 with residents. These interviews were conducted after analysing the survey data in order to confirm the emerging modes of life identified through statistical analysis. In 2024, I continued with approximately 30 interviews, focusing on the gendered division of labour among individuals engaged in environmentally conscious lifestyles. While green communities were the primary focus of my initial fieldwork, the extended duration of the research allowed me to capture their dynamic evolution.

References

Benoît, Y. & Tanquerelle, H. (2010). *La communauté: Entretiens [The commune: Interviews]*. Futuropolis.

Blanc, L. (2012/1839). *Organisation du travail* [*The organisation of work*]. Hachette Livre BNF.

Brown, S. L. (Ed.). (2002). *Intentional community: An anthropological perspective*. State University of New York Press.

Chiffoleau, Y., Akermann, G., Paturel, D., & Noël, J. (2023). Des circuits courts à la sécurité sociale de l'alimentation: Économies concrètes et récit politique pour la solidarité alimentaire [From short food supply chains to food social security: Material economies and a political narrative for food solidarity]. *Lien social et politiques* [*Social Ties and Politics*], 90, 310–329. https://doi.org/10.7202/1105099ar

Coulangeon, P., Demoli, Y., Ginsburger, M., & Petev, I. (2023). *La conversion écologique des Français: Contradictions et clivages* [*The ecological conversion of the French: Contradictions and cleavages*]. PUF.

Dubuisson-Quellier, S. (2010). From consumerism to the empowerment of consumers: The case of consumer-oriented movements in France. *Sustainability*, 2(7), 1849–1868. https://doi.org/10.3390/su2071849

Escribano, P., Lubbers, M. J., & Molina, J. L. (2020). A typology of ecological intentional communities: Environmental sustainability through subsistence and material reproduction. *Journal of Cleaner Production*, 266, Article 121803. https://doi.org/10.1016/j.jclepro.2020.121803

Gausset, Q., & Jensen, P. D. (2024). Living sustainably in a Danish eco-community: How social and physical infrastructures affect carbon footprints. *npj Climate Action*, 3(1), 33. https://doi.org/10.1038/s44168-024-00113-5

Graeber, D. (2010). Communism. In K. Hart, J.-L. Laville, & A. D. Cattani (Eds.), *The human economy* (1st ed., pp. 199–210). Polity Press.

Halfacree, K. (2022). Revisiting 1960s countercultural back-to-the-land migration and its millennial resurgence. *The Global Sixties*, 15(1–2), 43–78. https://doi.org/10.1080/27708888.2022.2133275

Hervieu-Léger, D., & Hervieu, B. (2005/1979). *Le retour à la nature: «Au fond de la forêt… l'État»* [*Back to nature: 'In the heart of the forest… the state'*]. Éditions de l'Aube.

Kingfisher, C. (2021). *Collaborative happiness: Building the good life in urban cohousing communities* (1st ed.). Berghahn Books. https://doi.org/10.3167/9781800732391

Kozeny, G. (1995). Intentional communities: Lifestyles based on ideals. https://articles-and-essays.s3.amazonaws.com/Intentional+community+/G-Kozeny-lifestyles-based-on-ideals.pdf

Langridge, N., Buchs, M., & Howard, N. (2023). An ecological basic income? Examining the ecological credentials of basic income through a review of selected pilot interventions. *Basic Income Studies* 18(1): 47–87. https://doi.org/10.1515/bis-2021-0044.

Leafe Christian, D. (2006). *Vivre autrement: Écovillages, communautés et cohabitats* [*Alternative ways of living: Ecovillages, communities, and cohousing*]. Ecosociété.

Lombard, P., & Baysse-Lainé, A. (2019). Terre de Liens, un levier foncier militant au service d'un projet politique pour l'agriculture [Terre de Liens: An activist land property regime in the service of a political project for agriculture]. *Économie rurale* [*Rural Economy]*, 369, 83–101. https://doi.org/10.4000/economierurale.7010

Magnusson, D. (2018). Going back to the roots: The fourth generation of Swedish eco-villages. *Scottish Geographical Journal*, 134(3–4), 122–140. https://doi.org/10.1080/14702541.2018.1465199

McCurdy, R. (2013). *Faire ensemble: Outils participatifs pour le collectif* [*Doing things together: Participatory tools for the collective*]. Passerelle Eco.

Meijering, L. (2012). Ideals and practices of European ecovillages. In M. Andreas, F. Wagner, & J. Seifert (Eds.), *Realizing utopia: Ecovillage endeavors and academic approaches* (pp. 31–42). Rachel Carson Center for Environment and Society (RCC).

Meijering, L., Huigen, P., & Van Hoven, B. (2007). Intentional communities in rural spaces. *Tijdschrift voor Economische en Sociale Geografie* [*Journal of Economic and Social Geography*], 98(1), 42–52. https://doi.org/10.1111/j.1467-9663.2007.00375.x

Metcalf, W. J. (1984). A classification of alternative lifestyle groups. *The Australian and New Zealand Journal of Sociology*, 20(1), 66–80. https://doi.org/10.1177/144078338402000105

Nielsen-Englyst, C., & Gausset, Q. (2024). From countercultural ecovillages to mainstream green neighbourhoods—A view on current trends in Denmark. *npj Climate Action*, 3(1), 60. https://doi.org/10.1038/s44168-024-00140-2

Pruvost, G. (2015). Chantiers participatifs, autogérés, collectifs: La politisation du moindre geste [Participatory, self-managed, collective worksites: Politicising the smallest gesture]. *Sociologie du travail* [*Sociology of Work*], 57(1), 81–103. https://doi.org/10.1016/j.soctra.2014.12.006

Pruvost, G. (2021). *Quotidien politique: Féminisme, écologie, subsistence* [*Politics of the everyday: Feminism, ecology, subsistence*]. La Découverte.

Pruvost, G. (2024). *La subsistance au quotidien: Conter ce qui compte* [*Everyday subsistence: Narrating what matters*]. La Découverte.

Rouvière, C. (2015). *Retourner à la terre: L'utopie néo-rurale en Ardèche depuis les années 1960* [*Back to the land: Neo-rural utopias in Ardèche since the 1960s*]. Presses Universitaires de Rennes.

Sallustio, M. (2022). *À la recherche de l'écologie temporelle: Vivre des temps libérés dans les collectifs néo-paysans autogérés. Une analyse anthropologique* [*In search of temporal ecology: Living liberated times in self-managed neo-peasant collectives. An anthropological analysis*]. Presses Universitaires de Rennes.

Sargent, L. T. (1994). The three faces of utopianism revisited. *Utopian Studies*, 5(1), 1–37.

Sargisson, L. (2012). Second-wave cohousing: A modern utopia? *Utopian Studies*, 23(1), 28–56. https://doi.org/10.5325/utopianstudies.23.1.0028

Snikersproge, I. (2022). Capitalism, subsistence farming and the (new) peasantries from the perspective of the French neorural movement. *Culture, Agriculture, Food and Environment*, 44(1), 53–62. https://doi.org/10.1111/cuag.12290

Snikersproge, I. (2023). Who are neorurals? Or, how capitalist time discipline dilutes political projects and makes it difficult to propose an alternative. *Economic Anthropology*, 10(1), 65–76. https://doi.org/10.1002/sea2.12258

Snikersproge, I. (2024). Ecological labour or why environmentally friendly practices struggle to become mainstream. *Ecological Economics*, 224, Article 108286. https://doi.org/10.1016/j.ecolecon.2024.108286

5

THE SOCIAL ARCHITECTURE OF CONSTRUCTIVE DISOBEDIENCE

Collective resistance in the Danish energy transition

Simon Lex

Introduction

In recent years, policy makers have identified 'green communities' to establish the legitimacy of an inclusive and sustainable societal development (Hoff, Gausset, & Lex 2020). By expanding environmental policies that primarily focus on individual consumption patterns, their goal is to develop social infrastructures that enable communities to collectively adopt more sustainable lifestyles (Gausset 2020). This strategy encourages people to take responsibility into their own hands by engaging in collective green initiatives such as regenerative agriculture, sustainable housing, recycling stations, shared transportation, renewable energy communities, and nature-based sanitation (Lex & Gausset 2022; see also introduction in this book).

In line with this political focus, a variety of community models have emerged in urban and rural areas in Denmark. There has been an increase in eco-villages, cohousing collectives, and green neighbourhood communities where people come together to share resources, promote sustainable living practices, and cultivate a sense of community through collaborative decision making, shared spaces, and environmentally friendly actions (Jarvis 2015; Hansen 2019). These intentional communities provide what Quentin Gausset and his colleagues have coined 'practivism' (Gausset 2020). They argue that this practical and collaborative approach strengthens the social relations among members and enhances their commitment to promoting a way of life that aligns with collective ideals and values (see also Gausset, Jensen, & Hunt 2019). Moreover, recent research suggests that practivism in green communities can significantly reduce carbon dioxide emissions (Gausset et al. 2025).

DOI: 10.4324/9781003769064-6

While the studies of green communities indicate a potential for advancing sustainable societal change, they have largely overlooked how civil concern and conflict unfold in everyday communal relations and interactions. This chapter contributes to the existing literature on the social architectures of practivism by demonstrating how alternative sustainable relations can arise from resistance. It focuses on civil groups that, while opposing top-down renewable energy projects, develop alternative solutions based on their lived experiences, local knowledge, and cultural and symbolic capital. Moreover, the chapter highlights that, in order to align local civil concerns with national climate policies, regional municipalities need to expand instrumental and individualised approaches and governance models. In other words, they should engage in and collaborate with local communities of 'experts by lived experience' in the planning and design of renewable energy projects. This reflects the concept of 'constructive disobedience', which critiques traditional disruptive activism and protest movements. In her book *No is Never Enough* (2017), Naomi Klein advocates for a more constructive approach to resistance; she emphasises the importance of pairing disobedience with the design of alternative policies and counterproposals.

In this chapter, with the help of anthropological fieldwork conducted in four locations in Denmark, I first explain how the development of large-scale solar and wind projects causes disruption and mistrust in related local communities. I demonstrate how the unequal distribution of costs and benefits, combined with a lack of transparency in the energy project development, results in disagreements and disputes among residents. I suggest that these disputes escalate from gossip to open conflicts that ultimately weaken social cohesion within the communities.

Second, I analyse how a group of protesters in Southern Jutland opposed a proposed renewable energy project planned for their region. The group of protesters not only voiced its dissatisfaction but also developed a counter-expertise that offered locally informed critiques of the project. The group went even further by designing an alternative proposal that aligned with the community's values and needs. Acknowledging this local input, the municipality decided to prioritise the perspectives of the group, which indicated a willingness to incorporate grassroots feedback in the decision making process.

Third, I suggest that by mobilising across projects and villages, local groups have evolved into a regional community of local voices that collectively engage in and respond to such renewable energy projects. I conclude that citizen groups, through their resistance and opposition, have become an influential collective actor in developing policy and decision making within the green energy transition in Denmark.

Before delving into the empirical and analytical findings, I first outline the context by discussing the overall energy transition in Denmark. I then introduce the theoretical framework, which includes the concepts of 'social communities', 'practivism', and 'constructive disobedience'. I finish this section by describing the method and approach applied in the study.

Context and theoretical framework

Renewable energy projects in Denmark

The decarbonisation of the energy sector in Denmark has led to the development of ambitious climate policies. The government has set a target to reduce carbon emissions by 70% by 2030 and aims for net-zero emissions by 2050 (Klima-, Energi- og Forsyningsministeriet [Danish Ministry of Climate, Energy and Utilities] 2022). To reach these targets, the government plans to quadruple onshore renewable energy production by 2030 and develop 32 large energy parks primarily in regions of Jutland and Sealand (Klima-, Energi- og Forsyningsministeriet 2023). While this political commitment indicates support for advancing the green energy transition and despite renewable energy production that reached record production levels in 2024, the development of large-scale onshore solar and wind parks continues to face challenges because of local concerns and resistance (Weisdorf, Vangkilde, & Lex 2025).

To address this issue and to secure community support—what has been defined as 'social acceptance' (Wüstenhagen, Wolsink, & Bürer 2007; Gade & Rowlands 2018)—the Danish government in 2024 increased compensation rates for residents living near the projects and allocated additional funding for community-building initiatives. Although these policies aim to reduce resistance, they reflect a commitment to centralised planning and financial incentives that neglect the important role of citizen engagement and dialogue in the application, development, and decision making processes of renewable energy projects.

Such a top-down approach to implementing large-scale, commercially driven renewable energy projects has faced criticism for undermining local agency and autonomy (Howe 2014; 2019). Scholars argue that this political strategy does not align with community values and priorities and therefore leads to a lack of trust in developers and government authorities (Devine-Wright 2005; 2009; Goedkoop & Devine-Wright 2016). According to a comment by Patrick Devine-Wright (2019), there has been a decreasing political emphasis on grassroots and citizen-led initiatives in the field of green energy. In what he terms 'Local Energy', he argues that policy makers have redirected their focus toward investments in large-scale green energy projects that are established and operated by private developers. He contends that this approach fails to recognise local knowledge and initiatives, which, according to research on community renewable energy, is instrumental in facilitating a socially and culturally sustainable transition (see Kirkegaard et al. 2021).

In response to the political emphasis on large-scale industrial investments and implementation, social movements and local communities have highlighted the need for a more inclusive approach in the development and decision making processes of renewable energy projects. They argue for greater shared ownership and increased citizen engagement to ensure a just transition. This perspective resonates with the concept 'community renewable energy' (CRE), which has been used to describe how people establish local energy projects collectively. The term was coined by Gordon Walker and Patrick Devine-Wright (2008) as a response to the 'Not-in-my-back-yard' (NIMBY) approach, which identifies civil resistance as a negative and individual response to external infrastructural interventions in local neighbourhoods (see Wolsink 2000; Batel 2020).

CRE focuses on the active engagement of civil society and communities, local decision making, shared ownership models, and a just distribution of benefits and harms (Walker & Devine-Wright 2008). The concept analyses the organisation of community renewable energy projects that follow a democratic decision making process that ensures inclusivity and transparency. In these practices, residents invest their time and financial resources in the development of the projects (Bauwens 2016; Seyfang et al. 2017). While CRE has offered valuable insights into project-driven initiatives, local decision making processes and shared ownership models over the past two decades, it fails to capture the informal collective actions that emerge in local communities as a response to the implementation of renewable energy projects. To develop a clearer understanding of how such actions influence local community building, I introduce the concepts of social community, practivism, and constructive disobedience in the following section.

Social communities, practivism, and constructive disobedience

A social community is characterised as a group of individuals who unite around common needs, interests, and values. Whether in physical or digital spaces, the members of the community engage in collective actions that lead to a sense of belonging (Lex & Hvenegaard 2020). In physical spaces, the community activities often unfold in local institutions, movements, or neighbourhoods, where face-to-face interactions reinforce social connections. Meanwhile, social communities flourish online on digital platforms that enable people from diverse geographical locations to connect, exchange experiences, and collaborate around shared interests.

From a conceptual perspective, it is difficult not to acknowledge Tönnies's distinction between '*Gemeinschaft*' and '*Gesellschaft*', which distinguishes between two types of social communities. *Gemeinschaft* (community) is typically associated with traditional, rural areas; it refers to close-knit, personal relationships that are characterised by emotional ties, shared values, and a strong sense of belonging. Conversely, *Gesellschaft* (society) is characteristic of modern, urban environments and describes more impersonal and formal interactions marked by rationality and individualism (Tönnies 1957 [1887]). More recently, Anthony Cohen (1985) contends that social communities are not only formed through interactions and relationships but also shaped by symbolic meanings that vary among members. This variability enables a diverse group of members to coexist under a shared identity.

Benedict Anderson's concept of 'imagined communities' enhances this discussion. Anderson emphasises the importance of shared narratives and collective identities that extend beyond personal relationships (Anderson 1983). He introduces the idea of 'deep, horizontal comradeship' and describes the nation-state as a diverse collective where individuals, through shared symbols and cultural practices, create a sense of belonging that does not depend on direct interactions (Anderson 1983: 6).

Anderson and Cohen focus on two critical aspects of community representation: whether tangible or imagined, communities are dialectically formed through shared narratives and symbols that are continuously reinforced through interactions with contemporary communities or social entities (Cohen 1985). Moreover, scholars often argue that the idea of community is defined by its boundaries, which help delineate a group's identity (Barth 1969). Cohen stresses the significance of these boundaries and notes that communities are relationally formed by differentiating themselves from other communities (Cohen 1985). Consequently, the concepts of identity, relationality, *and* differentiating are fundamental to understanding the notion of a social community.

In the context of social infrastructures of the green transition, it is relevant to introduce 'green intentional communities', which are composed of individuals who come together around shared values and goals. In these communities, members adopt a collaborative lifestyle that prioritises environmentally friendly practices. Examples include eco-villages, co housing collectives, and green neighbourhood initiatives, where participants work together to share resources, promote sustainable living practices, and cultivate a sense of community through joint decision making, shared spaces, and environmentally conscious actions (Jarvis 2015). In intentional communities, 'practivism' serves as a pragmatic approach to driving behavioural changes. It highlights the value of lived experiences, practical skills, and direct engagement in creating alternative solutions that challenge existing unsustainable systems and practices. Practivism addresses local needs and concerns and recognises the collective knowledge and expertise of community members. By emphasising bottom-up initiatives and day-to-day collective practices, it empowers individuals to utilise and mobilise their resources and skills with the aim of contributing to broader systemic changes (Gausset, Hoff, & Lex 2019).

Additionally, this pragmatic and constructive approach has been extended to more informal and even resistant collectives or movements. In this context, the concept of 'constructive disobedience' becomes relevant: it redefines traditional acts of resistance not merely as oppositional gestures but as proactive efforts to create space for alternative futures (Klein 2017). Constructive disobedience emphasises the importance of pairing dissent with well-developed counterproposals that present feasible pathways for change, thereby establishing a moral and ethical basis for opposing unjust policies or practices. This perspective suggests that resistance is most effective when it focuses on building new structures rather than dismantling existing ones. Thus, opposition can adopt a constructive stance that highlights the necessity of incorporating diverse citizen-driven perspectives and horizons (Weisdorf, Vangkilde, & Lex 2025).

Methods

My analysis is based on ethnographic fieldwork conducted over 24 months from 2022 to 2024. Initially, I developed a fieldwork plan that included three regional locations and municipalities: Bornholm, Esbjerg, and Fredericia. These municipalities were selected because they are involved in planning and implementing various small- and large-scale renewable energy projects. Bornholm and Esbjerg have also been politically identified as critical areas for advancing the green energy transition in Denmark.

The fieldwork was carried out by a research team consisting of three student assistants, two postdocs, and one senior researcher (the author). As the fieldwork evolved, we identified relevant opportunities for further inquiry, which led us to expand the data collection to include the regions of Aabenraa, Tønder, Skive, Kolding, and Thisted. In each of these sites, we identified and engaged with local citizens and municipalities and conducted interviews in their everyday private or professional environments. While the interviews focused on common themes related to citizen engagement, community building, and green energy transition, we developed specific interview guides for each participant based on their unique positions and perspectives.

Overall, our fieldwork resulted in 45 qualitative interviews with citizens living near proposed renewable energy projects across the identified regions. Additionally, we participated in meetings and conducted 22 individual and two group interviews with 26 employees from 22 municipalities. In addition to the eight municipalities initially mentioned, we also engaged with municipal workers from 14 other municipalities involved in the deployment of renewable energy projects.

During the fieldwork, we recorded and filmed semi-structured interviews and documented observations and informal conversations through detailed field notes. After returning from field visits, our research team systematically analysed the data collected from the interviews and participant observations. We then conducted two workshops to identify patterns within the dataset. This workshop provided a solid foundation for understanding local particularities at the field sites while facilitating comparative analysis across regional scales. Finally, we cross-validated and compared our results with existing literature by triangulating identified patterns and themes.

Empirical findings

Lack of transparency

In 2018, residents of a small village in Southern Jutland discovered that a solar panel park was planned for their neighbourhood. John, who had lived in the area his whole life, received the news while shopping for groceries in the local supermarket. A former friend and regional politician jokingly mentioned to him that he would likely soon have a new neighbour, meaning the solar park. Despite having heard rumours about other villages in the municipality, where large industrial energy projects were being developed without properly informing their potential neighbours, John was taken by surprise. He and his wife, Britta, knew most of the members of the village council and had strong connections with local politicians in the area. Still, they had heard nothing.

When John shared the news with Britta, he realised that the hopeful thoughts he had had on the ride home about how the new green energy park would provide a much-needed economic boost to the community were not at all shared by his wife. Instead, Britta voiced her scepticism about the potentially disruptive effects of the industrial park on their daily lives and the surrounding landscape. She was concerned that the construction of the site would alter their way of life and lead to noise and increased traffic. She was also afraid that the solar park would destroy the serene landscape of their neighbourhood and threaten the area's unique identity. Britta's sense of home and community represented the core values of the region, which was marked by expansive agricultural fields and a peaceful environment.

While John could see the potential for economic growth that the park could bring to the area, he also shared Britta's concerns. He understood that their home and community were defined by the quiet life they had grown to love over decades.

After weeks of seeking information about the project, the couple finally received an update from one of their acquaintances, who was at the city council. They learned that, indeed an industrial developer planned to construct a large-scale solar park just a few hundred metres from their property. This news angered Britta: she was adamant that external industries should not take over their community. As the weeks went by, Britta and John grew increasingly distrustful of the project and the people and institutions involved. Britta, who had moved to the area after marrying John in the 1970s, said that the lack of transparency felt like a slap in the face by outsiders who were ignorant of their lives and community. She felt excluded and uninformed, which she considered a sign of disrespect, especially towards John's family, who had lived in the area since the 19th century. She noted that through decades of diligent work in agriculture, his family had laid the foundation for their communal identity and cultural heritage. Their hard work and commitment over the years had contributed to preserving the traditions and values of the area and now, because of decisions made by politicians in Copenhagen, they were disregarded and treated as insignificant and invisible members of the community.

Internal disagreements

John, who had an inherent distrust of large city companies, was not surprised by the developers' behaviour. 'That's just business', he said. Rather, he felt disappointed by the lack of communication from long-time friends in the area, particularly their neighbour Toby, who owned the land designated for the solar park. John suspected that Toby had made a deal with the developer to lease or sell his land; it was rumoured to be a very profitable arrangement. John and Toby had known each other since childhood, and although they had lost touch over the past couple of decades, he did not understand why Toby had not reached out to him. Normally a mild-tempered man, John had to admit that he was angry and had no desire to engage with Toby and his family.

John and Britta were not the only ones caught in the dispute spurred by the development of the renewable energy project. Two years back, Phil, who for decades had been a successful farmer, was approached by an energy developer interested in renting his land for a solar panel park. Phil said that while he would prefer to continue his work as a conventional farmer, the past decade had been financially challenging. After his son showed no interest in taking over the farm, Phil began considering alternative means to support his family. He also noted that he was not getting any younger, which made the attractive offer from the developer even more appealing.

After Phil decided to accept the offer from the developer, all hell broke loose. Some neighbours expressed their anger during community meetings, while others chose to cut off communication with him and his wife. Phil explained that he understood his neighbours' concerns: the solar panels were not the most visually appealing investment. Nevertheless, he said that he had not interfered with what others chose to do with their land, and despite feeling isolated, he did not hold any grudge against his neighbours. He regarded his decision as a positive step because he had always been interested in pursuing a good business opportunity. Although he admitted that he had never previously invested in environmentally friendly farming, he said that renewable energy would have a positive sustainable impact on the area.

Another local resident, Lone, reported that developers had approached landowners individually with confidential contracts and instructed them to keep quiet and refrain from discussing the agreements with their neighbours. She believed that developers significantly contributed to local gossip and mistrust by selectively reaching out to some neighbours while excluding others, by secretly negotiating with individual landowners, and by pressuring the municipality to approve their plans without genuinely consulting with residents of the communities. Lone thought that this approach was a core cause of the local conflict. While she was in favour of the green energy transition, she concluded that the internal disagreements highlighted how the 'green energy rush' was tearing local communities apart.

An alternative strategy

In the year after discovering the plans for establishing the solar panel park, John and Britta explored the field of renewable energy. They researched publicly accessible information from other projects in the municipality and examined the effects of solar panels on both people and the environment. They reached out to Mariann and Ben, who lived near the proposed site and were also opposed to the project. Together, they wrote letters of criticism to the municipality, submitted responses during the hearing process, participated in public meetings to voice their objections to the development, and built alliances with other resistance groups throughout the region. Nevertheless, Britta noted, their efforts did not change the plans. This and other renewable energy parks were essential for the municipality to achieve climate targets and support green growth and new job opportunities in the region.

Despite their complaints being ignored, Britta and John refused to give up. Together with Mariann and Ben, they decided to adjust their strategy. Drawing on months of experience as protesters, the small group designed an alternative proposal for a solar park, which was smaller and located closer to a nearby busy road. As John pointed out, since the project was inevitable, they might as well attempt to guide it in a beneficial direction. He stressed that the group did not oppose renewable energy or the possibility of having such projects in their region. Instead, they wanted their voices to be heard and to have a say in how the project was designed and where it was situated. The group submitted their proposal, and in a surprising turn, the municipality decided to go with their plan. What started as resistance to a concrete solar panel park developed into the design of an alternative suggestion that was positively received by the municipality.

Interestingly, residents in other villages and communities in the area began to mobilise similarly and developed alternatives to the commercially driven development of the renewable energy parks. Eric, a community leader, said that the mobilisation was driven by concerns that industrial interventions would lead to ecological degradation and socio-economic injustices. By collaborating across various projects and villages, he said, and engaging with the municipality and energy developers, residents could meaningfully incorporate their perspectives and needs into project development. These efforts allowed them to advocate for practices that aligned with their values and priorities and thus reclaim some kind of agency over their environments and communities. Eric also said that the groups sought the expertise of a professional negotiator, which brought legal and administrative insights to the table and enhanced their ability to navigate complex negotiations and agreements. This local mobilisation amplified the group's collective influence and enabled them to become a more powerful actor in negotiations regarding the implementation of green energy parks and environmental policies and practices.

Thus, in addition to individualised resistance and push-back, the communities recognised the strength of unity. They mobilised across smaller villages and established a regional alliance that provided them with a stronger, collective voice. Their goal was to make it harder for developers to overlook their demands and concerns. By harnessing local knowledge and advocating for inclusive decision making, the communities played an important role in reshaping the narrative regarding the centrally governed green energy transition. They ensured that it aligned not only with technical solutions, but also with the lived experiences and aspirations of those most impacted by the changes: the local community. This form of collective action transcended mere opposition; it embodied an approach with active engagement based on local knowledge, practical skills, and lived experiences.

Analysis and discussion

Resistance as community building

I have outlined in the empirical descriptions how local communities perceive and engage with the development of renewable energy projects in Jutland, Denmark. These accounts serve as a lens for understanding how concern and resistance evolve into a social process of constructive disobedience and community mobilisation over time. While the introduction of renewable energy projects often incites conflict, mistrust, and local struggles among community members, municipality staff, and private developers, this study shows how concerns and disagreements can transform into constructive collective actions that help to protect local values and interests within the context of the green transition.

John and Britta's first encounter with the solar panel park proposal marked the beginning of years of struggle that was driven not only by concerns over aesthetic and environmental effects but also by disruptions to cultural identity and community values. Similarly, on a broader scale, the local protests were sparked by fears that the large parks would negatively affect the residents' lifestyles, harm their local environment and landscapes, and threaten their cultural identity. When learning about the renewable energy projects, residents began to articulate their concerns, which over time contributed to the formation of a shared narrative. For instance, Britta's feelings of exclusion and lack of transparency resonated with others who shared similar experiences of feeling ignored and marginalised by developers and local authorities. These concerns cultivated a sense of solidarity among the residents and enabled them to see their struggles as part of a shared cause rather than isolated injustices.

What began as a series of individual interactions, for instance, the ongoing informal encounters in Britta and John's home, evolved into more concerted efforts to strategically negotiate the terms of the renewable energy projects. By acknowledging the local values, building counter expertise, seeking strategic partnerships, and leveraging collective identities, the community transformed into a more influential actor in the negotiations with the municipality and energy developers. This process not only amplified a shared belonging and identity in the community, it also instilled a sense of common purpose to protect values and interests in a time of transition. This, I suggest, illustrates how resistance can serve as a component of community building. When residents encounter external pressures that threaten their traditional ways of life, such as the development of large industrial projects, economic changes, or policy interventions, they recognise the importance of collective action and local agency in preserving agricultural traditions, the natural landscape, and cultural identities. The experiences of Britta, John, and their neighbours exemplify how social communities form around common concerns and struggles. This aligns with Cohen's and Anderson's assertion that social communities are constructed through shared experiences and narratives that promote a shared sense of belonging and solidarity (Cohen 1985; Anderson 1983).

A practivist approach to renewable energy

During the collective resistance process, John and Britta met with other residents in their homes and in local community centres to exchange experiences, deepen their knowledge, and plan for future activities. This practical work enabled them to examine and assess the project's implications and to gain a deeper understanding of technical details, economic factors, legal frameworks, and their democratic rights. This research, if you will, gave them the capacity to engage in informed negotiations with private developers and regional authorities. The involvement of professional negotiators further enhanced their ability to navigate complex legal and economic issues. This suggests that combining local knowledge with expert insights can lead to more equitable outcomes.

In addition, John and Britta's interactions with other concerned residents transformed into alliances within the community. Over time, residents from surrounding villages also started to organise against similar renewable energy projects. By harnessing local knowledge, taking collective actions, and building regional alliances, these community groups challenged powerful actors who had previously marginalised local voices. This organising of regional alliances enabled them to confront the top-down and centrally governed energy projects.

This example from Southern Jutland demonstrates that concerned communities can not only be opponents of renewable energy policies and practices, they can also be catalysts for reshaping them. Instead of merely resisting passively, John, Britta, and their neighbours engaged in the political and practical development of the projects. Over time, by building expertise, knowledge, and networks, they gradually converted their initial frustrations into new proposals. Their constructive approach enabled them to move beyond simple opposition to a position of influence, where they could incorporate local knowledge and cultural context into the design of alternative renewable energy projects. As these residents developed a community of resistance, they not only expressed their concerns about perceived injustices but also actively contributed to (re)shaping a sustainable transition that respects the community's identity, values, and aspirations.

Collective disobedience

As previously noted, the lack of transparency and communication during the project development intensified feelings of exclusion and mistrust among the residents of the communities affected by the project. Importantly, the concerns raised by Britta and John extended beyond the solar park itself; they were acts to protect their community's identity, values, and traditions. Unlike purely oppositional or symbolic acts of protest, the resistance turned into engagement aimed at influencing policy and practice through informed participation. John, Britta, and others mobilised local knowledge, engaged in regional alliances, and developed alternative proposals.

Although the renewable energy projects initially created conflicts among the residents, the resistance movement gradually became a relational practice uniting local voices situated in community values and culture identity as well as in the national political landscape of green transitions. In this way, these experiences reflect Klein's concept of constructive disobedience (Klein 2017), which emphasises that dissent and resistance should be complemented by alternative visions and solutions. The experiences in Southern Jutland thus exemplify a community-led resistance that navigated and reshaped socio-political and environmental discourses.

This form of constructive disobedience was not an isolated act by members of one community. The region-wide mobilisation indicated a shift from individual and localised dissent to the emergence of a collective voice. As community members engaged in the green energy policy across villages, they drew upon their shared knowledge and experiences to confront centrally governed plans and projects. I suggest that the shift from individual concerns to collective negotiations enhanced the notion that constructive disobedience can serve as a means of empowerment and advocacy rather than simply being a reaction against external intrusions and interventions.

By advocating for increased participation in decision making processes, these communities exemplify how residents can elevate marginalised voices and promote a more democratic approach to the renewable energy transition. The findings I have discussed demonstrate that resistance, when framed through constructive disobedience, can stimulate engagement, community identity, and challenge the centrally controlled governance around renewable energy development. Through engagement and the cultivation of alliances, the residents did not merely resist the imposition of external projects. Motivated by their lived experiences, cultural practices, and a shared sense of community, they also demonstrated how engagement can transform the narrative around renewable energy. In doing so, they paved the way for a green energy future that is rooted in community empowerment, transparency, and participation.

Moreover, I believe that this chapter demonstrates how the residents, by taking responsibility into their own hands, formed local practivist communities over time. However, while existing literature underscores the constructive collective impacts of practivism, it often overlooks how tension, conflict, and civil concern shape everyday communal interactions—especially in contexts of opposition to top-down green projects. This chapter contributes to filling this gap by demonstrating that resistance not only serves as a form of protest but can also be a foundation for generating alternative solutions rooted in local knowledge, lived experiences, and cultural capital.

I also contend that regional governance must evolve beyond instrumental and individualised approaches to sustainable energy. Instead, local authorities should engage communities as 'experts by lived experience' in designing renewable energy projects that respect local concerns, practices, and knowledge systems. This shift towards participatory governance reflects a broader argument: resistance rooted in local agency and embedded in community practices can be transformed into productive negotiations that support more just energy transitions.

This chapter highlights the complex nature of practivist communities. It acknowledges that alternative sustainable practices often arise from opposition and civil concerns. This perspective contributes to a nuanced understanding of the social architectures that underpin a more inclusive green transition driven by communities themselves.

Conclusion

I have analysed how communities engage in the deployment of climate and energy policies in Denmark. The findings of this analysis indicate a deep local concern about the development of renewable energy projects such as solar panel and wind parks. Residents are worried about negative environmental impacts, loss of cultural identity, and disruption to their everyday lifestyles. Moreover, issues such as the unequal distribution of costs and benefits and a lack of transparency in the planning and implementation of the projects contribute to tensions and disagreements. These conflicts can escalate into disputes and even undermine the social cohesion in the communities.

However, the emergence of an organised and informed resistance exemplified by the group of protesters in Southern Jutland demonstrates a constructive approach to the renewable energy projects. Rather than simply voicing their discontent, these groups of residents collectively crafted alternative proposals that aligned with the values and needs of their community. The proposals encouraged the municipality to acknowledge and incorporate local perspectives in the decision making process. Thus, the local community played a vital role in shaping the deployment of the regional and climate energy policies.

The chapter also emphasised the capacity of civil groups to mobilise and collaborate across projects, villages, and municipalities. Through their collective efforts, the groups not only promoted more equitable and community-driven renewable energy projects but also evolved into an influential actor in the decision making processes of the regional green transition. Through concerted efforts to enhance social ties and community cohesion, the civic groups demonstrated how their actions, what I refer to as constructive disobedience, can serve as a catalyst for productive engagements and advocacy in local governance.

Such informal collective practices anchored in local values and knowledge not only strengthen the legitimacy of renewable energy projects but also illustrate how communities participate in designing alternative solutions to overcome the challenges associated with deploying renewable energy in Denmark.

References

Anderson, B. (1983). *Imagined communities*. Verso Books.
Barth, F. (1969). *Ethnic groups and boundaries: The social organization of culture difference*. Universitetsforlaget.
Batel, S. (2020). Research on the social acceptance of renewable energy technologies: Past, present and future. *Energy Research and Social Science*, 68, 1–5.
Bauwens, T. (2016). Explaining the diversity of motivations behind community renewable energy. *Energy Policy*, 93, 278–290.
Cohen, A. (1985). *Symbolic construction of community*. Routledge.
Devine-Wright, P. (2005). Beyond NIMBYism: Towards an integrated framework for understanding public perceptions of wind energy. *Wind Energy*, 8(2), 125–139.
Devine-Wright, P. (2009). Rethinking NIMBYism: The role of place attachment and place identity in explaining place-protective action. *Journal of Community & Applied Social Psychology*, 19(6), 426–441.
Devine-Wright, P. (2019). Community versus local energy in a context of climate emergency. *Nature Energy*, 4, 894–896.
Gaede, J., & Rowlands, I. H. (2018). Visualizing social acceptance research; A bibliometric review of the social acceptance literature for energy technology and fuels. *Energy Research and Social Science*, 40, 142–158.
Gausset, Q. (2020). Stronger together. How Danish environmental communities influence behavioural and societal changes. In: J. Hoff, Q. Gausset, & S. Lex (Eds.), *The role of non-state actors in the green transition: Building a sustainable future* (pp. 52–70). Routledge.
Gausset, Q., Suszkiewiecz, N. J., & Nielsen-Englyst, C. (2025). *Grønne fællesskaber og carbon fodaftryk* ('Green communities and carbon footprint'), pp. 1–40. SAMSKAB and Copenhagen University. https://doi.org/10.13140/RG.2.2.22464.60165
Gausset, Q., Jensen, K.H.B., & Hunt, J. B. (Eds.). (2019). *Vigør: Fortællinger fra den grønne frontlinje*. [Vigor: Stories from the green frontline]. Ingerfair.
Goedkoop, F., & Devine-Wright, P. (2016). Partnership or placation? The role of trust and justice in the shared ownership of renewable energy projects. *Energy Research Social Science*, 17, 135–146.
Hansen, A. H. (2019). Et bofællesskab og en bevægelse: En antropologisk analyse af økosamfundsaktivisme på tværs af skalaer [A co-housing community and a movement: An anthropological analysis of eco-village activism across scales]. *Politik*, 22(3), 99–114.
Hoff, J, Gausset, Q., & Lex, S. (2020). Introduction. In J. Hoff, Q. Gausset, & S. Lex (Eds.), *The role of non-state actors in the Green transition: Building a sustainable future* (pp. 1–13). Routledge.
Howe, C. (2014). Anthropocenic ecoauthority: The winds of Oaxaca. *Anthropological Quarterly*, 87(2), 381–404.
Howe, C. (2019). *Ecologics: Wind and power in the Anthropocene*. Duke University Press.
Jarvis, H. (2015). Towards a deeper understanding of the social architecture of cohousing: Evidence from the UK, USA and Australia. *Urban Research & Practice*, 8, 93–105.
Kirkegaard, J. K., Cronin, T., Nyborg, P., & Karnøe, P. (2021). Paradigm shift in Danish wind power: The (un)sustainable transformation of a sector. *Journal of Environmental Policy & Planning*, 23(1), 97–113.

Klein, N. (2017). *No is not enough: Resisting Trump's shock politics and winning the world we need.* Haymarket Books.
Klima-, Energi og Forsyningsministeriet [Danish Ministry of Climate, Energy and Utilities]. (2022). Klimaaftale om grøn strøm og varme 2022. Et grønnere og sikrere Danmark. Danmark kan mere II [Climate agreement on green electricity and heating 2022. A greener and safer Denmark. Denmark Can Do More II]. Regeringen. www.regeringen.dk/nyheder/2022/aftale-om-et-mere-groent-og-sikkert-danmark/
Klima-, Energi og Forsyningsministeriet [Danish Ministry of Climate, Energy and Utilities]. (2023). Aftale om større energiparker på land og mere kompensation til naboer til solceller og vindmøller [Agreement on larger onshore energy parks and increased compensation for neighbors of solar panels and wind turbines]. Regeringen. www.regeringen.dk/nyheder/2023/aftale-om-stoerre-energiparker-paa-land-og-mere-kompensation-til-naboer-til-solceller-og-vindmoeller/
Lex, S., & Gausset, Q. (2022). Fællesskab som drivkraft for den grønne omstilling [Community as a driving force for the green transition]. *Økonomi og Politik* [*Economy and Politics*].
Lex, S., & Mikkelsen, H. H. (2020). The stop wasting movement as a community of potentialities. In J. Hoff, Q. Gausset, & S. Lex (Eds.), *The role of non-state actors in the green transition: Building a sustainable future.* Routledge.
Seyfang, G., Hielscher, S., Hargreaves, T., Martiskainen, M., & Smith, A. (2014). A grassroots sustainable energy niche? Reflections on community energy in the UK. *Environmental Innovations and Societal Transitions*, 13, 21–44.
Tönnies, F. (1957 [1887]). *Community & society* [*Gemeinschaft und gesellschaft*]. Michigan State University Press.
Walker, G., & Devine-Wright, P. (2008). Community renewable energy: What should it mean? *Energy Policy*, 36(2), 497–500.
Weisdorf, M.,Vangkilde, K. T., & Lex, S. (2025). Diverging energy horizons. Rethinking public resistance in the Danish renewable energy transition. *Energy Research & Social Science*, 121. https://doi.org/10.1016/j.erss.2025.103945
Wolsink, M. (2000). Wind power and the NIMBY-myth: Institutional capacity and the limited significance of public support. *Renewable Energy*, 21(1), 49–64.
Wüstenhagen, R., Wolsink M., & Bürer, J. M. (2007). Social acceptance of renewable energy innovation: An introduction to the concept. *Energy Policy*, 35(5), 2683–2691.

6

APPLYING PARTICIPATORY METHODS OF GREEN COMMUNITIES TO TRANSFORM RURBAN AREAS OF NORTHWESTERN GERMANY

Anne-Kathrin Schwab

Introduction

Democracy is under threat, particularly in rural areas of Germany, where citizens are increasingly losing trust in established political systems and turning towards alternatives at the political fringes of society. In the face of multiple environmental and economic crises and the crisis of democracy (Schwab 2020) that involve structural changes, especially in rural and rurban areas, many people feel unheard and unable to meaningfully influence political decisions. Yet for decades, eco-villages and community-led initiatives (CLIs) such as Lebensgarten and Steyerberg and others in the Lower Saxony region (Schwab 2020) have been experimenting with participatory democratic models and building cultures of consensus- and consent-based decision making (cf. Kunze 2020; Andreas 2013; Schwab & Royssen 2022). These initiatives serve as living laboratories for social innovations (Kunze 2012) and raise the crucial question: how can participatory methods proven in eco-villages be adapted for use in broader, more traditional communities (cf. Veciana et al. 2020)?

This chapter addresses this question by discussing an action research project conducted from 2022–2025 in the rural village of Dunum in Northwestern Germany. The project was developed within the framework of the '4N: Nordwest Niedersachsen Nachhaltig Neu' ('Northwest Lower Saxony: Sustainable and New') project funded by the Lower Saxony Ministry of Science and Culture. Within the 'subproject 5' (https://vier-n.de/project/teilvorhaben-55/) of this project, the University of Vechta, which is located approximately 80 miles south of Dunum, explored strategies for participatory sustainability transformations in rurban areas.

DOI: 10.4324/9781003769064-7

The term 'rurban', used in the sub-project 5 of the project 4N 'Rurban Design and Mobility' is a neologism and specifies the tension field between rural and urban (cf. Schwab et al. 2024). It defines the polarities of a line between the two 'poles' of rural and urban since most spatial definitions cannot be defined as totally rural anymore. The term is also based on the transformative idea of making urban areas more rural, e.g., urban gardening projects and other production of ecological products so that cities do not need to rely solely on rural extraction and supply Revi et al. 2006).

As part of this initiative, Dunum served as a case study to experiment with methods adapted from eco-village practices and to explore how these can be integrated into the traditional social structures of a rural village. The central goal of this chapter is to illustrate one possible avenue for 'greening' an existing community using participatory action research. In the current context, where sustainability moves beyond eco-communities and into the mainstream of society, Dunum provides an instructive example.

Sustainability here is not treated as an exclusively environmental issue but as fundamentally social: without a functioning, participatory community, there is no social energy to drive ecological transformation. By carefully weaving participatory techniques into the cultural fabric of Dunum while respecting its traditional neighbourhood structures and informal governance patterns, this case study shows that it is possible to foster collective action even in a setting far from the original context of eco-villages. However, the project also reveals the challenges inherent in this process, including structural barriers, power dynamics, and historical divisions within the community. This paper demonstrates how participatory approaches from the eco-village movement can inform rural sustainability transitions. It proposes that fostering a socially cohesive and participatory community is a necessary precondition for environmental sustainability, which is a relevant lesson for the green transformation of rural Europe far beyond Dunum.

Transformative science, action research, and prefigurative experiments

The need for profound societal change in the face of climate, biodiversity, and social crises challenges traditional forms of science. In response, transformative science has emerged as a new paradigm. It redefines the relationship between science and society by no longer confining science to an observational role and recognising its active part in shaping sustainable futures (Schneidewind & Singer-Brodowski 2013; Singer-Brodowski, Holst, & Goller 2021).

Transformative science is deeply normative. It explicitly acknowledges that societal transformations such as the current 'Great Transformation' movement toward environmental sustainability (WBGU 2011) are not automatic processes, but can and must be actively shaped. Scientific institutions, researchers, and educational structures are called upon to become change agents that contribute to systemic shifts through research, education, and institutional reform. In this view, science must take responsibility not only for knowledge production but also for its societal consequences (Vogt 2022; Vogt & Weber 2020).

This theoretical framework situates my project in Dunum. The aim is not simply to study rural transformation but to initiate a participatory transition process towards a more sustainable, socially cohesive rural community. To operationalise these goals, the research draws upon action research, an approach grounded in participation, collaboration, and iterative learning cycles (Kemmis et al. 2014; Bradbury 2015). Action research aligns naturally with transformative science by:

- Actively involving local citizens as co-researchers;
- Working toward practical improvements and empowerment;
- Integrating cycles of planning, acting, observing, and reflecting; and
- Combining scientific inquiry with real-world experimentation.

Following Wittmayer and Schäpke (2014), action research not only produces knowledge about social change but becomes part of the change itself. It embodies the core principles of transformative science: embeddedness, reflexivity, normativity, and responsibility.

The methodological and conceptual foundations of this project are inspired by practices of eco-villages and the Transition Town movement. Eco-villages have long served as experimental fields, real-world laboratories, or living labs, for participatory democracy, collective sustainability practices, and alternative economies (Bergaglio et al. 2025; Kunze 2012; Schwab 2020; Veciana et al. 2020). A real-world laboratory is defined as format for transformative sustainability research co-created by citizen and scientists where societal impact is being experimented and measured (cf. Parodi et al 2021). These communities prefigure desired societal futures through everyday practices: consensus-based decision making, shared ownership models, regenerative agriculture, renewable energy cooperatives, and community-based social support systems. In the vocabulary of transformative science, these communities enact prefigurative politics: they embody in the here and now the social, ecological, and economic relations they seek in order to generalise at broader scales (Schneidewind 2018). They offer concrete experiences of 'living alternatives', not only by envisioning sustainable futures but by testing their feasibility in practice.

Similarly, the Transition Town movement, originating with Rob Hopkins (2008), demonstrates how local communities can mobilise collective action to build resilience and sustainability even beyond the confines of intentional communities. It emphasises localised food systems, renewable energy, community-building, and participatory governance. This project adapts selected methods and practices from movements (including Dragon Dreaming, Theory U, sociocracy-inspired decision structures, visioning workshops, and real-world sustainability experiments) to the specific context of a traditional rural village such as Dunum, which had not previously undergone such systemic experimentation.

Thus, the research in Dunum serves as an attempt to bridge the experimental cultures of eco-villages and Transition Towns with the lived realities of traditional rural communities. It addresses a key challenge identified by transformative science scholars: moving sustainability transitions from niche experiments into the mainstream of society (Grunwald 2018; Schäpke et al. 2018). One of the core outcomes of this research is the finding that sustainable transformation in rural and rurban contexts is most effective when it begins by strengthening the social fabric, i.e., building trust, fostering inclusive participation, and enabling local empowerment. Only when these social foundations are in place do broader environmental and economic changes gain traction and legitimacy within communities. Without a functioning community, there is no basis for effective sustainability transitions. I present here the methodology used to design and implement this participatory transformation process in Dunum and ground it in the principles of transformative science and action research.

Living the research: Ethnography, participation, and co-creation in Dunum

My research approach is deeply rooted in transformative science; it combines Grounded Theory, qualitative action research, and ethnographic immersion. In this project, I became a co-learner and shaped change together with others. Instead of studying the village of Dunum as an external outsider, I became part of its everyday life: I lived in the village, attended local events, and made sure I was approachable and present in the most natural ways. This closeness allowed me to build trust, engage in authentic dialogues with residents across all generations, and understand the complexities of the community from within.

My research journey started with my presenting the project to the mayor and some interested citizens, followed by extensive qualitative interviews that I analysed through Grounded Theory (Glaser & Strauss 1998) and continuous participant observation. I facilitated participatory workshops inspired by Dragon Dreaming (Croft 2011) and Theory U (Scharmer 2022), methods that focus on co-creation, shared leadership, and egalitarian participation that have also been adopted by the eco-village and Transition Town movements for use in their programmes. In line with the principles of transformative research, I did not position myself as an external expert, but rather as a learner and catalyst walking alongside the villagers in a shared process of reflection and action.

Ethnography and grounded theory

After I presented my *bona fides* to Dunum's social and political leaders, I began to make myself known in the community, began the empirical work, and entered a phase of intensive participant observation and ethnographic immersion. Living with my child for extended periods in a caravan in Dunum and accompanied by my horses and a dog, I became part of everyday village life. This natural presence fostered low-threshold contact, especially with farmers and rural residents, and blurred the line between researcher and community. Daily walks and participation in local events enabled trust-based conversations and deep insights into everyday life, informal power structures, and social dynamics that are aligned with the principles of collaboration and mutual learning of action research.

I also conducted semi-structured interviews with key local stakeholders: long-term residents, council members, farmers, and newcomers. Using Grounded Theory (Glaser & Strauss 1998; Charmaz 2006), I analysed the research material inductively to allow local perspectives to shape the understanding of the special-structural transformation dynamics.

Participatory workshops: Co-creating transformation and ethical reflection

Building on interviews and observations, I facilitated several participatory workshops with around 50 participants each, which is a notable portion of Dunum's population. People were extraordinarily enthusiastic about sharing their opinions and their views. Drawing on methods from Dragon Dreaming and Theory U (Scharmer 2016), the workshops emphasised co-creation, deep listening, shared responsibility, and emergent planning. To ensure inclusive participation, I used culturally familiar language that was developed with the locals (e.g., using the word *dorfschnack* instead of 'communication practices') and accessible visual tools like World Café and thematic tables. I also incorporated the Global Ecovillage Network's 'Map of Regeneration' (GEN International)[1] to connect local ideas with global sustainability goals. Outcomes included the identification of priority action areas and the formation of self-organised working groups that involved both citizens and council members.

Throughout this process, I maintained strong ethical reflexivity and clearly communicated my dual role as researcher and facilitator. I navigated insider-outsider tensions by staying neutral in local conflicts and suggested the hiring of an external moderator when needed. My everyday presence—often accompanied by my child and dog—unintentionally lowered social barriers: my human and canine companions made me more approachable and helped to foster trust. In line with transformative science, I made my normative stance explicit: I was promoting social, ecological, and cultural sustainability (Schneidewind & Singer-Brodowski 2013; Grunwald 2018; Vogt 2022). This helped create an open, collaborative atmosphere for mutual learning.

Results: Transformation and structural change in Dunum

Historical transformation and structural change: Insights from interview analysis

The interview analysis that I conducted reveals that Dunum, like many rural communities in Germany, has undergone profound structural transformation over the past 60 years. What was once a vibrant, agriculturally anchored, and socially self-sufficient village has become a sparsely populated, car-dependent commuter community with weakened local structures.

Historically, Dunum's economy was shaped by small-scale farms and additional rural businesses such as pubs, shops, and craft enterprises that sustained both the economy and the social life of the community (Kiesè & Böök 2010). However, political and economic pressures have led to the steady decline of small and mid-sized farms in Dunum and the region surrounding it. These pressures included structural agricultural reforms such as joining separately-owned agricultural land together to be able to use bigger machines in order to produce more output, European Union (EU) regulations, and market dynamics favouring large-scale production. Only a few large agricultural enterprises remain and parallel businesses have vanished, resulting in not only a loss of informal meeting spaces but also the greater loss of social cohesion.

A major tension identified in the interviews with villagers concerns the farmers' ambivalent relationship with green and sustainability policies. Farmers are especially critical of ecological goals because they feel economically pressured even without the added expense of following ecological farming methods. In addition, while EU and regional frameworks demand environmentally sustainable practices, globalised markets reward quantity over quality. As one long-time farmer put it: 'They ask us to farm greener, but they pay us for quantity, not for quality. It's impossible to live like that without getting bigger or giving up' (Interview, 15 June 2024).

With the collapse of agricultural diversity and the closure of local businesses, most residents work in nearby towns such as Esens and Wittmund. This shift has reshaped Dunum's daily life: people commute by car, shop in supermarkets like Aldi and Combi, and spend less time in their own village. Walking and cycling have declined; public spaces often appear empty. This reduces spontaneous encounters and weakens the village's social life. As one young mother observed: 'Here, you don't meet people walking or cycling around like in the city. Everyone just gets into their car and drives away' (Workshop discussion, 17 July 2024).

Social structures and community dynamics in Dunum

Despite the economic and infrastructural challenges Dunum faces, the village still possesses a strong and deeply rooted social infrastructure. This community fabric is primarily organised around neighbourhood groups, which play a central role for Dunum's residents that lasts a lifetime.

Dunum residents belong to geographically defined neighbourhoods, which are tightly knit social groups based on spatial proximity rather than on kinship (Interview, 6 April 2024). These neighbourhoods support people 'from cradle to grave': they celebrate life events such as births, weddings, and funerals through shared rituals like *Kränze binden* (wreathmaking) that reinforce intergenerational bonds. Changing one's home address within the village also changes one's neighbourhood affiliation, which highlights the territorial nature of social belonging.

Beginning about ten years ago, more people began to move to Dunum. This was caused by a number of factors, including the community initiative 'Dunum und umzu' ('Dunum and around it'), which demonstrated the application of transformative and innovative approaches in more rural areas. In addition, Dunum is close to the North Sea but is not overcrowded with tourists, as are Bensersiel or other villages that are even closer to the sea. Furthermore, Dunum has a primary school, which makes it especially attractive for families with younger children (Interview, 3 May 2024).

With the increasing number of newcomers, Dunum actively works to integrate them into local traditions. Rituals like the North German tea ceremony (*Teeritual*) are central to this process. Held daily at fixed times, the ceremony follows a specific etiquette: tea is poured by the hostess, followed by fresh cream and a lump of *Klüntje* sugar that crackles in the cup (Interview, 20 August 2024). During the tea, the life-long villagers gently correct any missteps made by newcomers to this ritual, which is a highly stylised opportunity to learn the village's social codes. I experienced this ritual as a newcomer during my fieldwork.

However, such strong social cohesion can present challenges. The aforementioned 'Dunum und umzu' initiative was organised in 2018 (before the Covid pandemic) by newcomers who were committed to applying sustainability goals to both the governing and the daily lives of the village. The initiative failed in large part because it did not anchor itself in existing neighbourhood structures and did not involve the local council enough. Despite the organisers' good intentions, it was perceived as a conflicting parallel structure to local authorities and decision makers and was firmly resisted by the local council and also by some villagers, whereas others were very sad when it stopped. Most of the initiative's projects faded, except for a 'Bücherzelle und Geschenke-Ecke' (low-threshold book and exchange hut) near the Hayungshof and the church (Interview, 25 July 2024). This illustrates a key lesson for transformative engagement: new initiatives must respect and work within traditional social frameworks, such as the communal elected leaders, in order to gain acceptance and sustainability.

Economic structure and mobility infrastructure

Dunum's shift from a locally anchored village to a commuter community has deeply changed the village's mobility patterns. Interviews and observations show that car use now dominates the village's modes of transportation: walking, cycling, and even using horses (for agricultural or mobility reasons) have sharply declined. With local jobs and shops gone, nearly all working-age residents commute by car to work and shop in nearby towns. Public transport is sparse and unreliable, and cycling paths serve the tourists more than they serve the residents' daily needs. As a result, even short trips within the village to visit neighbours or attend events often require a car. One resident put it bluntly: 'Without a car, you are cut off'.

This car dependency has reshaped Dunum's social life: streets are often empty during the day and informal encounters while walking or cycling have become rare. Also, cars speeding loudly through the village can be dangerous for both bicycling children and pedestrians. Younger generations now grow up viewing cars as essential, which reinforces the trend towards owning cars and increases the relevance of cars for their egos. As a senior resident recalled: 'We used to meet people on the way to the shops, now people disappear inside their cars'.

The workshops and action research of our project identified key barriers to sustainable mobility: there is a lack of infrastructure for walking and cycling and current cultural norms favour cars as the default mode of transportation. Dunum's economic priorities are focused on road maintenance rather than local solutions to the problems of implementing local business ideas and establishing infrastructure.

At the beginning of the research project, there was some interest in change among the villagers. Residents attending the workshops tentatively suggested ideas such as shared mobility (e.g., village taxis and ride-sharing) and safer walking and cycling routes (especially for school children) to places like the Hayungshof, the church and school, and other community-based transport solutions. However, any mobility efforts must align with the local economic conditions as well as Dunum's neighbourhood structures and social norms in order to be accepted and effective. Accordingly, a focus on sustainable mobility should be embedded within broader local economic and social solutions. In Dunum, improved mobility is therefore less a starting point than a follow-up to other sustainability projects and processes, as mobility responds to existing needs and functions primarily as a supportive bypass rather than a driver of change.

Workshops and action research approaches

The workshops in Dunum were carefully based on participatory methodologies from Dragon Dreaming and Theory U, combined with approaches that are commonly used in eco-village settings. Special emphasis was placed on:

- *Inclusivity*: Ensuring that all voices regardless of age, gender, social status, or length of residence are heard and valued equally;
- *Cultural sensitivity*: Integrating local traditions and dialects into the workshop language and topics to create a sense of familiarity and belonging; and
- *Visual and dialogue tools*: Using World Café methods, poster presentations, and hands-on group exercises to facilitate participation beyond verbal discussion.

The turnout for the workshops exceeded expectations. The first workshop on 6 July 2024 was attended by approximately 50 residents. A follow-up workshop on 17 July 2024 again saw over 50 participants, including many young families, teenagers, and elders.

Key topics and collective priorities

During the workshops, residents were asked to sit around tables to encourage discussion. They first reflected on the findings from the interviews before moving into thematic discussions.

The following key themes emerged from these workshop discussions:

- *Creation of a vibrant village centre* ('the *Thing*'[2]): A strong desire was expressed for a central meeting point—a village café, pub, or shop—where people could meet informally, connect, and strengthen community life. Participants emphasised that face-to-face encounters are vital for maintaining Dunum's social fabric.
- *Family-friendly Dunum* (*Familienfreundliches Dunum*): Young parents and grandparents proposed creating a new playground and regular intergenerational events to foster bonds between different age groups. The

idea of a 'Café for Young and Old' that would host quarterly community gatherings at the Hayungshof also emerged from these discussions.

- *Improving communication and access to information* (*Dorfschnack*[3]): Participants recognised the need for better information sharing about local events, council decisions, and association activities. Suggestions included revitalising the village website, enhancing information displays around the village, and publishing regular community calendars.
- *Mobility and accessibility* (*Hin und her*[4]): As outlined in the previous section, the community voiced concerns about car dependency and brainstormed ideas for mobility hubs, bike paths, and car-sharing initiatives. However, residents acknowledged that such changes would require sustained effort and external support.
- *Renewable energy and sustainability* (*erneuerbare energien und nachhaltigkeit*): While more controversial, renewable energy projects (such as community solar installations) were also discussed; participants showed both an interest in but also scepticism about how local residents would benefit directly from new developments.
- *Addressing conflicts and preserving tradition*: Lastly, a table dedicated to 'grievances and conflicts' collected frustrations about unresolved disputes in the village, a lack of transparency in local government, and fears that Dunum would lose its traditional values amidst modernisation efforts. Residents emphasised that future initiatives must respect Dunum's deep-rooted customs, such as neighbourhood structures and collective rituals.

Formation of working groups

Following the workshops, three main working groups were established; each was supported by council members and volunteers from the community.

The working groups were:

- The Village Centre Working Group focused on creating a social hub and investigating possibilities for a multifunctional café or meeting space;
- The Family-Friendly Dunum Group concentrated on planning and organising the Café for Young and Old, developed playground proposals, and fostered activities for young families; and
- The Information and Communication Group dedicated itself to updating and expanding the village's communication channels, including digital and physical noticeboards.

Smaller thematic groups related to mobility and renewable energies were also organised and are expected to become more active in the future as the process matures.

Reflections on the participatory process

After conducting the participatory workshops in Dunum, several key observations emerged regarding how the process unfolded. The following section reflects on the participatory approach that we used. It examines how unfamiliar facilitation methods were adopted by the villagers, how conflicts were navigated, and how villagers were empowered to create a shared vision of the future. These reflections provide insights into the effectiveness of participatory action research in a traditional rural village context and highlight the importance of cultural sensitivity, continuity, trust-building, and empowerment.

Adoption of eco-village and Transition Town movements' methods

One striking aspect of the project was the introduction of participatory methods drawn from the eco-village and Transition Town movements. The facilitator introduced several innovative group techniques to the Dunum project, including sitting in a circle without tables in front of the participants, elements of Dragon Dreaming or Theory U exercises (Scharmer 2018), and sociocratic decision making elements, which are commonly used in intentional sustainable communities.

For example, Dragon Dreaming is a holistic community project process with phases of dreaming, planning, doing, and celebrating (Croft 2011). Theory U (Scharmer 2016), on the other hand, blends systems thinking and mindfulness to help groups sense and realise emerging possibilities for a sustainable and just future within their community. Similarly, sociocracy (dynamic governance) introduces inclusive, circle-based decision making with consent, which ensures that every voice is respected.

These approaches were entirely new to the villagers, whose prior experience was mostly with conventional town hall meetings and top-down planning. Initial reactions to these methods were mixed. Many participants were puzzled, even sceptical, about activities like a 'systemic mirror' presentation or a positioning exercise that involved moving around the room. Such formats were outside the community's comfort zone: after the second workshop, a few villagers privately commented that these methods felt 'a bit strange and ambiguous'.

In early sessions, participation in the more unconventional exercises was hesitant. For instance, when asked to physically position themselves towards the Map of Regeneration of GEN (Global Ecovillage Network), some people hung back, unsure if they should take it seriously. One participant in the second workshop said, 'This is very unconventional for people in Dunum. We are different. We don`t do things like that, we are just different than other villages'. The facilitator addressed this reluctance by patiently explaining the purpose of each exercise in down-to-earth terms and by linking them to local issues and reflecting back to the meta-level of sustainability.

Over time, acceptance grew as the villagers began to see the value of the new methods. As the workshops progressed, even the sceptics grew more comfortable and curious. Participants observed that these techniques actually helped them communicate and think together in new ways. For example, a reflective dialogue inspired by Theory U allowed individuals to share personal observations without interruption, which led to deeper understanding among neighbours. Likewise, using sociocratic-style rounds (speaking in turns) ensured that the quieter residents had a chance to contribute, which in turn fostered the creating of circles in order to make decisions in small groups. The group eventually came to appreciate it as a very effective interactive tool.

By the later sessions, there was a noticeable shift in the mood and tenor of the discussions: there was laughter in the room and creative ideas emerged more spontaneously during activities that initially had felt awkward. In one session, a normally reserved farmer engaged enthusiastically in a role-playing game about future village scenarios—something unimaginable at the start of the workshop. This evolution from resistance to engagement underscored an important point: with patience and practice, innovative participatory methods can take root even in a traditional community once people experience their benefits. Indeed, the group eventually began to experience the magic of working together as a group towards a common goal through these methods, which certainly validated the facilitator's eclectic approach.

Handling conflict through broad participation

At the outset, a significant social challenge threatened the process: a long-standing conflict between two prominent men in the village. This feud had dominated community meetings and had created a polarising atmosphere at village meetings for years; villagers tended to take sides or avoid meetings altogether to escape the tension. In the participatory workshops, the researcher consciously chose not to acknowledge this conflict. Instead of attempting direct mediation between the two rivals (which might have further spotlighted their power struggle), the action researcher designed a process that diluted the antagonists' influence by engaging a broader range of villagers. Workshops actively invited diverse participants, not only the usual outspoken residents, but also women, young people, and elders who had not been part of the earlier conflicts. Activities were structured in small mixed groups so that no single individual could dominate the discussion in plenary.

This broader engagement strategy had a noticeable effect: with many voices in the mix, the two men's dispute no longer took centre stage. They were present and participated, but their rivalry was overwhelmed by the community's collective focus on the topics at hand. For instance, in a visioning discussion about improving local public spaces, attention shifted to the ideas coming from a range of people (a teenager's suggestion for a sports event, a mother's idea for a playgroup, an elder's memories of past festivals) rather than getting stuck on the points of contention between the two men.

By shifting the focus to inclusive participation, the workshops transformed the social dynamics. The previously dominant conflict was essentially sidelined, not by ignoring it, but by crowding it out with constructive dialogue involving many stakeholders. Participants later noted that meetings felt 'refreshingly productive' (notes from the observant participation at the workshop) compared to past gatherings because the feud could not derail the agenda. In effect, the conflict lost much of its momentum and its audience. This experience echoes the notion that proactive community engagement can preempt or diffuse conflicts by fostering wider ownership of the conversation. As more residents found their voice and collaborated on various initiatives, the feud between the two men gradually receded into the background. Once the men who saw that the community was moving forward, they became less inclined to clash in public.

The Dunum case illustrates how broadening participation served as an informal conflict resolution mechanism: it reduced polarisation by creating space for new alliances and topics to discuss, which allowed the village to direct its attention toward common goals rather than personal grievances.

Empowerment and co-creation of initiatives

A core principle of the Dunum process was to empower local residents as co-creators of change, rather than position them as passive participants. The researcher deliberately adopted the role of facilitator by guiding the process and providing tools but not actively influencing the content of the discussions. In practical terms, this meant the workshops were designed to encourage villagers to moderate their own working groups, define the issues that mattered to them, and devise solutions together. Community members were treated as equal partners in knowledge generation. This approach aligns with the participatory action research ideal that those directly affected by an issue are best positioned to understand their own needs and craft relevant solutions. In the workshops, this commitment to empowerment was evident in how the responsibility for the meetings was shared.

For example, during a session on improving village amenities, the facilitator opened the floor for volunteers to lead breakout groups on sub-topics like 'family activities' and 'renewable energy', topics that were directly relevant to the participants. Several town council members and even some initially reticent residents stepped up to guide these small-group discussions. The researcher intentionally delegated roles so that ultimately the locals were in charge of driving the conversations. This created a sense of shared leadership. For instance, two council members were asked to co-moderate the workshop series along with the researcher, rather than the researcher being the sole authority figure. The effect of this co-creative setup was empowering. Villagers began to see that the power to make change lay in their own hands. As they facilitated dialogues and saw their neighbours respond to their delegation, their confidence grew. Thus they saw that no one from the outside (like a researcher or facilitator) could make such changes. Rather, they could do it by themselves, by creating space that others can push things forward and creating spaces themselves where self-efficiency by themselves and by others were experienced. This was for all sides: the council members and the villagers an outstanding experience.

One villager asked the researcher, 'And who can we talk to, who is taking the responsibility (*wer trägt den Hut* = 'wearing the hat'[5]) for this? Is it you'? The researcher answered, 'No, but it could be you, if you dare, any one of you can wear the hat'. When he answered, 'But the hat doesn't suit me', the researcher replied, 'Oh, I think it would suit you very well'. Councillors told me later that this particular person is usually very disruptive in meetings and critical of every opinion or proposal. Since the researcher as an outsider was not aware of this, she unconsciously took away from this participant his habit of using his negativity as a weapon.

This empowerment was not just felt individually, it was also felt collectively: the community began to recognise its own agency. Instead of expecting an external expert or the government to come in and fix things, residents started taking initiative. After the formal workshops ended, the working groups continued meeting on their own without needing the researcher's presence. Locals organising themselves is a strong indicator of increased capacity and ownership. It resonates with the notion that community engagement empowers residents to shape decisions that affect their lives. By co-designing and moderating their own projects, Dunum's residents moved from being subjects of change to authors of their future; in doing so, they reflected the genuine empowerment aim of the participatory action research approach.

Dreaming and visioning exercises

Among the various methods used, the 'dream circle' visioning exercise (drawn from the Dragon Dreaming toolkit) was especially impactful. In one of the early workshops, participants formed a circle and were invited to articulate their personal dreams and wishes for the future of Dunum. The atmosphere was informal and encouraging: everyone had an equal chance to speak and all ideas, no matter how simple or grand, were welcome.

This exercise tapped into the imaginative and emotional side of community planning and allowed people to express what truly mattered to them for their village. Crucially, the process was structured so that each person shared in turn without interruption, and all contributions were recorded on a big poster that symbolised a collective vision in the making. People of all ages attended: children, youth, adults, parents, and elderly people had the same right to speak about their wishes and dreams.

The results were surprising in how grounded they were. Far from being daydreams of unrealistic projects, most people's visions centred on revitalising familiar aspects of village life. Many dreams reached into Dunum's history and identity. For example, one elder recalled the old inn that used to be a lively meeting place and envisioned it reopened as a community café. A young mother dreamed of bringing back an annual festival in early May to raise the maypole[6] (*Maibaum aufstellen*) that had faded away as a way to celebrate local culture and produce.

Others talked about very practical improvements: a playground, a shared space for young mothers, a bike trail, and a pedestrian route around Dunum for tourists. What stood out about these improvements was that these aspirations were tangible and achievable. The workshop participants weren't asking for high-tech smart city solutions or huge investments; they were dreaming of renewing the social and physical infrastructure the village already had or once had. This grounded nature of the dreams likely made them more resonant: people could immediately picture these things happening because many were part of living memory.

Sharing these dreams had a powerful psychological effect on the group. As each person spoke, heads nodded around the circle; participants discovered that they had overlapping hopes and values. The collective energy in the room shifted towards optimism as the villagers realised that they all wanted many of the same good things, as one participant put it. By the end of the dream circle, a rough shared vision had emerged organically from the mosaic of individual dreams—a vision of a more lively, connected, and self-reliant Dunum.

According to the Dragon Dreaming philosophy, creating such a shared vision is a critical first step that builds commitment and trust. The exercise in Dunum did this and more: it boosted the sense of unity and possibility among participants. Several people later mentioned that hearing others speak from the heart made them feel more united. It also strengthened their sense of agency; because the dreams were concrete and came from the community, people felt these are 'our' ideas, and we can work on them together.

In essence, the dream circle helped transform vague hopes into a collective vision with momentum. It instilled a renewed hope that the village's future is not predetermined by outside forces or past conflicts; instead, the future is something the villagers could actively co-create. This aligns with the broader Transition Towns approach, which sees communities as the drivers of envisioning and enacting a positive future, encouraging proactive participation and responsibility in shaping their own destiny. Equally important, the dreaming exercise was followed up by planning steps (consistent with the Dragon Dreaming method where dreaming is only the first phase). Because the visions were down-to-earth, the group could immediately start discussing how to realise some of them. In this way, the visioning directly fed into action, turning hope into pragmatic next steps. The experience of collectively imagining and then planning for their dreams significantly bolstered the villagers' confidence. It showed them that no idea was merely wishful thinking; with collective effort, even cherished dreams like these could move toward reality.

Overall reflections and lessons learned

Reflecting on the participatory process in Dunum, a number of broad insights emerge about conducting transformative action research in a traditional rural setting. First and foremost, the case demonstrates the importance of cultural sensitivity and patience when introducing new participatory practices. The eco-village-inspired methods eventually proved effective, but only because they were facilitated with care for the local context. An outsider suddenly coming in with flipcharts and facilitation games could have easily backfired. In Dunum, success hinged on building understanding and trust step-by-step: explaining unfamiliar exercises, respecting local communication styles, and showing how these methods served the village's own goals. In other words, external tools were adapted to local culture; they were not imposed. This allowed villagers to gradually embrace novel ways of collaborating without feeling that their traditions were being overridden.

Secondly, broad and inclusive engagement was a key to unlocking community momentum. By ensuring that a diverse cross-section of residents were involved from the beginning, the process reduced the usual hierarchies and gatekeeping that often exist in small communities. This had several benefits: it mitigated the influence of prior conflicts and elite actors, as discussed above, and it brought in fresh energy from people who hadn't been active before. The Dunum experience confirms that when more people have a stake and a voice, the community develops a stronger collective drive. Importantly, early inclusion of many voices also built legitimacy for the outcomes: since ideas came from various corners of the village, there was a sense of shared responsibility for implementing them. This finding echoes general principles of equitable engagement: when community members see that their voices are heard and respected, it fosters buy-in and stronger partner relationships and reduces friction down the line.

Third, empowerment through co-creation proved to be both a means and an end. The participatory process not only aimed to produce plans for village improvement, but also to build local capacity and confidence along the way. By consciously stepping back and pushing local actors to the forefront, the action researcher helped cultivate new local leaders and skills. This empowerment is a lasting outcome: long after the external facilitator leaves, the community now has individuals who have practiced facilitation, project planning, and collaborative problem-solving.

In a rural village that previously had limited experience with such approaches, this development of human and social capital is invaluable. It increases the likelihood that the community can continue the transformative work on its own. Indeed, continuity was observed in Dunum because some initiatives that persisted and new meetings were organised by the villagers themselves after the project had ended. Such continuity is a crucial success factor: it is the difference between a one-off experiment and a sustained transformation. The Dunum case illustrates that when people own the process, they are more likely to carry it forward. This aligns with the idea that early involvement and shared ownership lead to more successful and sustainable initiatives because the community has the motivation and know-how to keep up the momentum.

Another overarching lesson is the centrality of trust-building in participatory rural projects. In Dunum, trust was not a given, it had to be earned through transparency, consistent engagement, and delivering on small promises (like sharing workshop results with everyone, following up on ideas, etc.). The researcher's role in building trust was critical: her being present in the village beyond official meetings, listening to concerns, and demonstrating respect for local knowledge all contributed to a trusting relationship.

Over time, villagers went from wary to willing, and eventually to full partners in the project. This trust enabled open dialogue: people felt safe voicing dissenting opinions or 'crazy' ideas without fear, which only enriched the process. Trust is somewhat intangible, but the Dunum case shows it is the bedrock of effective participation. Without it, tools and techniques alone would have fallen flat. With trust, however, even bold visions and complex collaborations became possible because participants believed in each other and in the facilitator's good intentions. The experience reinforces the idea that trust and rapport are fundamental in action research, especially in close-knit communities where outsider initiatives will be scrutinised carefully.

The participatory process in Dunum also illustrates how transformative action research can catalyse change even in a small, traditional village when it is approached thoughtfully. By blending outside inspiration with local wisdom, managing conflicts through inclusion, empowering community co-leadership, and fostering a shared vision, the process helped unlock the community's latent potential. Dunum's journey was not without challenges, but the reflections from this case affirm that when given the opportunity, rural communities can be enthusiastic architects of their own sustainable future. Cultural sensitivity, broad participation, empowerment, visioning, trust, and continuity create the conditions for genuine transformation.

In the humble village of Dunum, these principles converged to turn scepticism into enthusiasm, conflict into cooperation, and distant hopes into actionable plans. Such outcomes provide a hopeful example for practitioners and researchers in transformative rural development: meaningful change is attainable when people are genuinely invited to participate, dream, and co-create the future of their community.

Discussion

The project in Dunum provides a compelling case study for the practical application of transformative research and action research principles in a traditional rural village. Transformative research as conceptualised by Schneidewind and Singer-Brodowski (2013) and further elaborated by Singer-Brodowski, Holst, & Goller (2021) emphasises not only the co-production of knowledge but also the co-creation of societal change in the direction of sustainability. This approach frames research as an embedded, participatory process that acknowledges the interdependence of science and society and aims to act as a catalyst for transformation rather than act as a distant observer (WBGU 2011).

Action research has long highlighted the importance of engagement, empowerment, and reflexivity (Kemmis et al. 2014). In Dunum, the research process aligned with these principles: villagers were engaged not as passive research subjects but as active partners who shaped the research questions, analysis, and outcomes. Reflexivity was embedded not only in the researcher's continuously adapting to local conditions but also by encouraging the community itself to reflect on its structures, practices, and potential future. The project thereby operationalised a modest and reflexive model of transformative science (Grunwald 2018), which maintains scientific rigor while openly addressing its own normative assumptions (Vogt 2022). It also demonstrated that action research processes can thrive in rural settings provided that trust-building, cultural sensitivity, and long-term engagement are prioritised.

A notable innovation of this project was the adaptation of methods derived from the ecovillage and Transition Town movements. These approaches, particularly Theory U, Dragon Dreaming, and sociocratic governance models are rarely applied in traditional villages. Prior research on eco-villages (Litfin 2014; Lockyer 2017; Schwab 2020) emphasises that such communities act as 'living laboratories' for experimenting with sustainable, participatory practices. However, scaling or adapting these models beyond self-selecting communities has been a notable challenge (Trainer 2019). Dunum's experience shows that partial and context-sensitive adaptation is possible even in non-intentional communities with deeply ingrained traditional structures.

Three key factors enabled this adaptation:

- Translation of practices into familiar frames: explaining participatory methods in practical, down-to-earth terms helped bridge initial scepticism;
- Respect for local culture: traditional rituals (such as the tea ceremony) and social structures (the neighbourhood networks) were not disrupted but honoured, which created trust; and
- Gradual introduction: complex or unfamiliar exercises were introduced incrementally, with ample space left for local adjustments.

Thus, this case study supports emerging scholarship that suggests that rural sustainability transitions can benefit from 'soft' adaptations of eco-village innovations (Kirwan et al. 2015; Kunze 2012). However, it also underlines the importance of local ownership: methods must empower villagers to act from their own cultural and emotional starting points rather than impose external blueprints.

The participatory strategy used in Dunum exemplifies key principles discussed in the literature on community empowerment and rural development (Pretty 1996; Chambers 1997; Bryden & Hart 2004). It resonates in particular with findings that successful rural engagement:

- Must involve a broad cross-section of the population in order to avoid elite capture;
- Must build on local knowledge and traditions rather than replacing them; and
- Must focus not only on planning projects but also on building local capacities and leadership.

By designing workshops that enabled villagers to dream, plan, and act collaboratively, the Dunum project fostered exactly these forms of empowerment. Importantly, conflict management occurred organically through inclusion rather than through confrontation, which illustrates that building broader engagement can sometimes be more effective than direct conflict mediation (Aiken 2017).

This experience suggests that even in small, aging rural communities, a significant latent social capital remains (Putnam 2000) that can be activated through thoughtful and participatory processes. It also indicates that sociocracy and other decentralised governance models can offer promising avenues for organising community action in rural settings, a finding that is increasingly discussed in rural transition studies (Kunze & Avelino 2015).

Despite the overall, undeniable success of the Dunum project, we observed several challenges:

- Initial resistance: Traditional villagers were sceptical of unfamiliar methods and needed time to develop trust;
- Sustainability of engagement: Maintaining momentum beyond initial workshops requires continuous support and empowerment; and
- Conflicts and rivalries: While partially mitigated, existing personal conflicts continued to influence dynamics subtly. These challenges echo critical voices in the debate on transformative research, who caution against romanticising participation (Strohschneider 2014; Gawel et al. 2017). Participation alone is not a panacea; it requires careful facilitation, attention to power dynamics, and long-term commitment.

Conclusion and outlook

The Dunum project demonstrates that transformative action research inspired by eco-village and Transition Town methods can foster meaningful change even in traditional rural communities. By combining trust-building, inclusive participation, culturally sensitive facilitation, and empowerment strategies, it was possible to catalyse community reflection, the co-creation of visions, and practical steps toward sustainability. The research contributes to the broader debate on transformative science by showing that adaptations of participatory methods originally developed in intentional communities can be viable in traditional contexts provided they are introduced respectfully and flexibly. Furthermore, the Dunum case highlights that empowerment through co-creation and attention to local social infrastructures such as neighbourhood networks and cultural rituals are crucial for sustainable outcomes.

At the same time, the case reaffirms that transformative rural development is a long-term process that must grapple with historical conflicts, demographic shifts, and structural economic challenges. Research and facilitation must embrace complexity, ambiguity, and ongoing reflexivity. Participation is not a guarantee of success; it must be carefully managed, embedded in existing social fabrics, and continuously supported through empowerment and capacity building. Ultimately, the experience in Dunum underscores a hopeful but pragmatic message: rural villages possess the social and cultural resources to navigate their own sustainable transitions if they are supported by participatory, respectful, and empowering processes. When practiced in this spirit, transformative science can truly become a catalyst for a more resilient and connected rural future.

Drawing from the experience in Dunum and connecting it with the state of the art in transformative and participatory research, the recommendations for implementing successful research projects are:

- Cultural sensitivity first: initiatives must deeply respect and engage with local traditions, rituals, and informal social structures. Rather than bypassing them, transformative approaches should build bridges to these traditions;
- Gradual introduction of innovative methods: participatory tools such as Theory U, Dragon Dreaming, and Sociocracy can be successfully introduced even in traditional settings if done incrementally and with clear, accessible explanations;
- Empowerment of local leaders: researchers and facilitators should focus on enabling local leadership rather than directing community processes themselves. Holding the space, rather than controlling it, proved essential in Dunum;
- Focus on collective, visible wins: supporting small, early successes (such as the establishment of meeting points or local initiatives) can build momentum and trust for larger transformations;
- Sustain engagement over time: working groups and participatory structures must be continuously nurtured through coaching, facilitation, and potentially external support to maintain energy and avoid fatigue;
- Acknowledge and navigate conflicts carefully: conflicts are a natural part of community life. Embracing pluralism and focusing attention on collective engagement can defuse tensions and broaden ownership; and
- Integrate sustainable practices early: introducing sustainability-related topics (e.g., local food initiatives and renewable energy) as part of broader community visions can make sustainability feel organic and relevant rather than abstract or externally imposed.

By implementing these recommendations, future transformative research initiatives— especially in rural settings—can enhance their capacity to not only study but also actively contribute to community-led transitions toward resilience and sustainability.

Notes

1 The Global Ecovillage Network (GEN), https://ecovillage.org/ecovillages/map-of-regeneration/ is an international association of communities and individuals dedicated to sustainable living practices that aims to restore ecological balance and foster social resilience. GEN facilitates the exchange of ideas, technologies, and cultural practices among its members to promote regenerative development worldwide. One of its key tools is the 'Map of Regeneration', an interactive

framework comprising 32 eco-village principles organised into five areas of regeneration: social, cultural, ecological, economic, and whole systems design. This map serves as both a diagnostic and educational resource to guide communities in assessing and enhancing their sustainability practices across multiple dimensions.

2 'The *Thing*' in Northwest Lower Saxony refers to a contemporary participatory format inspired by historical assemblies in Germanic legal traditions. In practice, it designates inclusive village forums such as *Dorfplenum* (village plenary) or gatherings in the *Dorfplatz* (village square) where residents deliberate and make decisions on local sustainability, infrastructure, and communal life. These assemblies reflect a revival of historic Germanic grassroots democratic practices within the framework of regional transformation and participatory governance.

3 *Dorfschnack* refers to an informal exchange of local information and news that is framed deliberately in Low German (*Plattdeutsch*) to resonate closely with the everyday language and cultural context of rural inhabitants in Northwest Germany. As part of the participatory research process, the term was used to explore local information flows and to identify which topics are truly relevant to village residents. Its linguistic anchoring in *Plattdeutsch* served to increase accessibility, authenticity, and trust among participants.

4 The term *Hin und her*, which is a phrase in Low German, means 'back and forth'. It was deliberately chosen for use in our research project in consultation with the local village council in order to frame discussions around rural mobility. Unlike the abstract and technocratic connotation of the term 'mobility', *Hin und her* reflects everyday language and lived experience. It captures not only the physical movement of people but also the social, emotional, and infrastructural dimensions of being on the move in rural life, which made it a more resonant and contextually grounded concept for local engagement.

5 'Wearing a hat' (*sich den Hut aufsetzen*) is a German idiom and means: to take responsibility, often voluntarily, for a task, role, or process, especially when it is unclear who should do it or when others hesitate. It implies that someone steps up, takes ownership, becomes the point person whom others can approach, and is accountable for how things move forward: 'Nobody wanted to do it, so I took responsibility'.

6 'Raise the maypole' (*Maibaum aufstellen*) refers to a traditional communal event in which a decorated tree or wooden pole (the maypole) is ceremonially raised on 1 May, often accompanied by food, music, and dancing. It is a community celebration that also requires a lot of collective effort.

References

Aiken, G. T. (2017). The politics of community: Togetherness, transition and post-politics. *Environment and Planning A*, 49(10), 2383–2401.

Andreas, M. (2013). Must utopia be an island? Positioning an ecovillage within its region. *Social Sciences Directory*, 2(4), 9–18.

Bergaglio, M., Capocefalo, V., Dal Borgo, A. G., & Gambazza, G. (2025). Ecovillages as living labs for social innovation: The case of Torri Superiore. *Sustainability*, 18(1), 188.

Bradbury, H. (Ed.). (2015). *The Sage handbook of action research*. Sage.

Bryden, J. M., & Hart, K. (Eds.). (2004). *A new approach to rural development in Europe: Germany, Greece, Scotland, and Sweden*. E. Mellen Press.
Chambers, R. (1997). Responsible well-being—A personal agenda for development. *World Development*, 25(11), 1743–1754.
Charmaz, K. (2006). *Constructing grounded theory: A practical guide through qualitative analysis*. Sage.
Croft, J. (2011). Dragon dreaming: Eine partizipative methode für die entwicklung gemeinschaftsorientierter nachhaltiger projekte [A participate method for the development of communal sustainable projects]. *Zeitschrift Bewusstseinswissenschaften. Transpersonale Psychologie und Psychotherapie* [*Journal of Consciousness Sciences. Transpersonal Psychology and Psychotherapy*], 17(1), 69–79.
Glaser, B. G., & Strauss, A. L. (1998). *Grounded theory. Strategien qualitativer forschung* [Strategies of qualitative research]. 4th edition. Verlag Hans Huber.
Gawel, E., Strunz, S., & Lehmann, P. (2017). Support policies for renewables: Instrument choice and instrument change from a public choice perspective. In D. Arent, C. Arndt, M. Miller, F. Tarp, & O. Zinaman (Eds.), *The political economy of clean energy transitions* (pp. 80–99). Oxford University Press.
Grunwald, K. (2018). Management sozialwirtschaftlicher organisationen zwischen steuerungsskepsis, dilemmatamanagement und postheroischer führung [Management of social economy organizations between skepticism about control, dilemma management, and post-heroic leadership]. In K. Grunwald (Ed.), *Sozialwirtschaft* [*Social economy*] (pp. 369–390). Nomos.
Hopkins, R. (2008). *The transition handbook: From oil dependency to local resilience*. Bloomsbury Publishing.
Kemmis, S., McTaggart, R., & Nixon, R. (2014). Introducing critical participatory action research. In S. Kemmis, R. McTaggart, & R. Nixon (Eds.), *The action research planner: Doing critical participatory action research* (pp. 1–31). Springer.
Kiese, D., & Böök, L. (2010). *Ortschronik Dunum und Brill* [*Chronicle of Dunum and Brill*]. Brune-Mettker Druck.
Kirwan, S., Dawney, L., & Brigstocke, J. (2015). *Space, power and the commons*. Taylor & Francis.
Kunze, I. (2012). Social innovations for communal and ecological living: Lessons from sustainability research and observations in intentional communities. *Communal Societies*, 32(1), 39–55.
Kunze, I., & Avelino, F. (2015). *Transformative social innovation narrative of the Global Ecovillage Network*. Report. TRANSIT project (Transformative social innovation theory). www.academia.edu/15251014/Kunze_I_and_Avelino_F_2015_Transformative_social_innovation_narrative_of_the_Global_Ecovillage_Network_TRANSIT_EU_SSH_2013_3_2_1_Grant_agreement_no_613169
Kunze, I. (2020). 'A new we': Post-individualistic community-based initiatives as social innovations? Empirical observations in intentional communities. In B. Jansen (Ed.), *Rethinking community through transdisciplinary research* (pp. 285–305). Springer International Publishing.
Litfin, K. T. (2014). *Ecovillages: Lessons for sustainable community*. Wiley.
Lockyer, J. (2017). Community, commons and degrowth at Dancing Rabbit Ecovillage. *Journal of Political Ecology*, 24(1), 519–542.

Parodi, O., Ober, S., Lah, O., Steglich, A., Wagner, F., & Podann, A. (2021). Herausforderung Reallabor: Werkstattbericht zur Reallaborforschung [The Real-World Lab Challenge: A Workshop Report on Real-World Lab Research.] *GAIA-Ecological Perspectives for Science and Society*, 30(4), 286–288.

Pretty, J. N. (1996). Participation, learning and sustainability: Emerging challenges for agricultural development. *Social Change*, 26(1), 7–33.

Putnam, R. D. (2000). *Bowling alone: The collapse and revival of American community*. Simon & Schuster.

Revi, A., Prakash, S., Mehrotra, R., Bhat, G. K., Gupta, K., & Gore, R. (2006). Goa 2100: The transition to a sustainable RUrban design. *Environment and Urbanization*, 18(1), 51–65.

Schäpke, N., Stelzer, F., Caniglia, G., Bergmann, M., Wanner, M., Singer-Brodowski, M., & Lang, D. J. (2018). Jointly experimenting for transformation? Shaping real-world laboratories by comparing them. *GAIA-Ecological Perspectives for Science and Society*, 27(1), 85–96.

Scharmer, C. O. (2016). *Theory U: Leading from the future as it emerges*. Berrett-Koehler Publishers.

Scharmer, C. O. (2018). *The essentials of Theory U: Core principles and applications*. Berrett-Koehler Publishers.

Scharmer, C. O. (2022). *Essentials der Theorie U: Grundprinzipien und anwendungen* [*Essentials of Theory U: Basic principles and applications*]. Carl-Auer Verlag.

Schneidewind, U. (2018). *Die große transformation: Eine einführung in die kunst gesellschaftlichen wandels* [*The great transformation: An introduction to the art of social chance*]. S. Fischer.

Schneidewind, U., & Singer-Brodowski, M. (2013). *Transformative wissenschaft. Klimawandel im deutschen wissenschafts-und hochschulsystem* [*Transformative science. Climate change in the German science and higher education system*]. Metropolis.

Schwab, A-K. (2020). *Transformation im ländlichen Raum. Ein Ökodorf und seine Wirkung in der Region* [*Transformation in rural areas. An ecovillage and its impact in the region*]. Springer.

Schwab, A.-K., & Roysen, R. (2022). Ecovillages and other community-led initiatives as experiences of climate action. *Climate Action*, 1(1), 1–4.

Schwab, A.-K., Yilmaz, A., Rolfes, D., Uphoff, S., Wolff, A., Robbers, L., & Fischer, F. (2024). Rurban design und mobilität [Rurban design and mobility]. In B. Schmidt, S. Steinigeweg, R. Pesch, C. Aka, B. Breckling, K. Baumann, & A.-K. Schwab (Eds.), *Reallabore in marsch, moor, geest und mee(h)r* [*Real-world laboratories in marsh, moor, geest, sea and more*]. *Transformation Dynamics* (pp. 149–162). University of Vechta.

Singer-Brodowski, M., Holst, J., & Goller, A. (2021). Transformative wissenschaft [Transformative science]. In T. Schmohl, P. Thorsten, & J. Schabert (Eds.), *Handbuch transdisziplinäre didaktik* [*Handbook of transdisciplinary didactics*] (pp. 347–356). Transcript.

Strohschneider, P. (2014). Zur politik der transformativen wissenschaft. Die verfassung des politischen [On the politics of transformative science. The constitution of the political]. In A. Brodocz, D. Herrmann, R. Schmidt, D. Schulz, & J. Wessel Schulze (Eds.), *Festschrift für Hans Vorländer* [*Commemorative publication for Hans Vorländer*] (pp. 175–192). Springer.

Trainer, T. (2019). Remaking settlements for sustainability: The simpler way. *Journal of Political Ecology*, 26(1), 202–223.

Veciana, S., Urbain, H., & Schwab, A-K. (2020). *Leben in zukunftsfähigen dörfern: Ökodörfer als katalysatoren nachhaltiger entwicklung* [*Living in sustainable villages: Eco-villages as catalysts for sustainable development*]. Umweltbundesamt.

Vogt, M. (2022). Enlightenment 2.0: Toward responsible science in the Anthropocene. In P. A. Wilderer, M. Grambow, M. Molls, & K. Oexle (Eds.), *Strategies for sustainability of the earth system* (pp. 395–409). Springer.

Vogt, M., & Weber, C. (2020). The role of universities in a sustainable society. Why value-free research is neither possible nor desirable. *Sustainability*, 12(7), 2811.

WBGU. Wissenschaftlicher Beirat der Bundesregierung Globale Umweltveränderungen [Scientific Advisory Board of the German Federal Government on Global Environmental Change]. (2011). Hauptgutachten welt im wandel gesellschaftsvertrag für eine große transformation [A changing world – A social contract for a great transformation.] Main report. www.wbgu.de/fileadmin/user_upload/wbgu/publikationen/hauptgutachten/hg2011/pdf/wbgu_jg2011.pdf

Wittmayer, J. M., & Schäpke, N. (2014). Action, research and participation: Roles of researchers in sustainability transitions. *Sustainability Science*, 9, 483–496.

7

THE PRACTICES OF SHARING

Negotiating spaces and things in three intentional community case studies

Penny Clark

Introduction

Decreasing the greenhouse gas emissions (GHGs) of residential housing will play a vital role in mitigating climate change and meeting UK Government net zero targets. In particular, energy use for heating and cooling homes, along with the indirect emissions associated with household goods, play a significant role: energy use in homes accounts for 16% of GHGs in the UK (BEIS 2023), and indirect emissions from goods account for approximately one-third of the total UK household's carbon footprint (Clark 2021).[1]

The UK Government's strategy has been primarily focused in recent years on innovating and implementing technology (e.g., heat pumps) rather than challenging domestic norms or practices. While technological change is essential, it has also proven to be extremely challenging to implement change at the necessary speed in order to meet targets because of the complexity and cost of retrofitting the UK's housing stock. The UK Department for Business, Energy and Industrial Strategy acknowledges that there is 'no silver bullet solution' (House of Commons Business, Energy and Industrial Strategy Committee 2022: 7); rather, many responses are required. One element that is currently under-utilised that could have significant impacts is behaviour change (Climate Change Committee 2020).

Given the challenges of technological transformation to decarbonise homes, behaviour change, or *practice transitions* (the more appropriate term for this discussion), presents an opportunity for rapid change, which may also have deeper and longer-lasting cultural implications for how we view, and act in, our homes and neighbourhoods. Intentional communities, which can be defined as 'a group of people who have chosen to live together or

DOI: 10.4324/9781003769064-8

share resources on the basis of common values' (Foundation for Intentional Communities n.d.), are valuable examples of domestic practices in which sharing spaces, resources, practices, social time, and values with people beyond the nuclear family unit is the norm.

Quantitative evidence shows the environmental benefits of shared households. Diana Ivanova and Milena Buchs (2020) found that the average carbon footprint of a household of five or more people has been found to be about half that of an average one-person household, while Anthony Underwood and Sammy Zahran (2015) found that by cohabitating, a person would, on average, reduce their carbon footprint by 23% when compared to their living alone.

Studies which specifically look at forms of intentional community also yield positive results. A literature review of quantitative environmental studies of cohousing communities and ecovillages found that the vast majority of studies 'strongly indicates that these intentional communities do achieve lower environmental impacts than comparable mainstream communities' (Daly 2017: 1369). The fieldwork upon which this paper is based found that the participating communities had GHGs that were between 37 and 67% of that of the average UK household, with significant emissions savings coming from energy use (associated with heating) and indirect emissions attached to household goods (Clark 2021).

Studies of intentional communities have detailed some of the practices and physical objects which lower environmental impacts, such as frequent use of renewable energy technologies and energy efficient design, food growing and consumption practices, travel practices, waste practices (Daly 2017), shared objects and amenities (Meltzer 2005), and smaller units with less floorspace per occupant (Bamford 2001; McCamant & Durrett 2011; Williams 2005). However, there is little available information about the activities, competences, and meanings around sharing spaces and things and what challenges and negotiations people face when doing so.

This chapter focuses on the sharing of spaces and things within intentional communities through an exploration of three UK intentional community case studies. By exploring these lived experiences, we find insights into the practical and social mechanisms of sharing spaces and things, which ultimately offers a means to lower domestic environmental impacts.

Materials and methods

This chapter is the result of both qualitative and quantitative research methods with three intentional community case studies (see Clark 2021). It focuses on qualitative outputs, although it occasionally refers to the quantitative measurement of the case studies' GHGs when this information is helpful for contextualising the qualitative data.

The intentional community case studies[2]

The case studies consist of two cohousing[3] communities and one house share community. This variety in community composition was chosen with a view to exploring the impact of spatial design and the number of people on the sharing of spaces and things in order to tease out variations and commonalities and gain a more holistic understanding of these practices, especially given the variety of intentional community typologies. The information about the ICs presented here was accurate as of the time of my fieldwork between October 2018 and January 2020.

Bramley Yard

At the time of my fieldwork approximately six years ago as of writing, Bramley Yard was a cohousing community with approximately 40 adults and ten children. Units were owner-occupied, and all residents were also members of the housing association. Residents were a mixture of retirees and those who worked, couples, single people, and families. The adults tended to be over 40, and all were White British and university educated. The building was a large, historic manor house which had been converted into 22 self-contained owner-occupied homes, and was situated rurally. Communal areas included two communal kitchens, flexible social spaces, workspaces, 40 acres of land on which the community collectively operated a farm, which they use to produce much of their own food. Residents were primarily brought together through their shared interest in running the farm. They also shared a general interest in living sustainably.

Meadow Cohousing

Meadow Cohousing had approximately 50 people living in 20 households. The tenure, or conditions under which buildings are held or occupied, was owner-occupied, with an ownership structure that was intended to preserve the affordability of the units in perpetuity. Residents included a mix of couples with young children and couples/single people who were either retired or working. Compared with Bramley Yard, the homeowner demographic of Meadow Cohousing was generally slightly younger, and a majority of residents were university educated. There was some ethnic diversity, though residents were mainly White. Their properties spanned ¾ of a hectare of land, with 20 homes that ranged from one-bed flats to four-bed houses. There was a common house with a kitchen and flexible meeting space, a meeting room, a bathroom and an office, allotments, and a 'pocket park' that was open to the public. The group expressed shared values of environmental sustainability, housing affordability, and mutual support.

Valley Rise

Valley Rise was a rented house share with six residents (three women, three men), four of whom were couples in their thirties and early forties. Residents were all university educated and almost all White. Their occupations included three psychotherapists, a civil engineer, a mental health worker, and a graduate student. The building was a four-bedroom, four-story semi-detached Victorian house. The shared space consisted of one bathroom and an additional toilet, a kitchen, a garden, and a large living area, which could be separated into two living areas via a folding door. Residents were brought together through their interest in meditation. They had a shared ethos of living together as a community and living sustainably.

Research methods

This fieldwork took place between October 2018 and January 2020. Methods included participant and non-participant observation, semi-structured interviews, focus groups, and document and image analysis. Observation took place during visits ranging from two weeks to a half-day. Photographs were taken or copies were made of materials containing written information, and photographs were also taken of communal spaces. Findings were written down at ad hoc periods during observation.

A total of 29 participants took part in semi-structured interviews with an average length of just over an hour. A total of 26% of Bramley Yard and Meadow Cohousing residents and 33% of Valley Rise residents were interviewed. For Meadow Cohousing, this percentage consisted of two out of six residents, which is acknowledged as a limited sample given the small size of the household. The data was analysed using a deductive and inductive approach drawing upon grounded theory and a theoretical framework drawing on social practice theory and social norms. Data was coded using qualitative analysis software NVivo, going through three rounds of coding, allowing themes and a theoretical framework to emerge.

Theoretical approach

This analysis adopts a theoretical approach which brings together social practice theory (SPT) and social norms.

SPT is a way of understanding human action that focuses on the action rather than the person performing that action (who is often referred to as an 'agent'). Actions are referred to as 'practices', which can be conceived as collectively shared sayings and doings held together as coordinated entities (Gram-Hanssen 2011). There have been various categorisations of what constitutes a practice (see for examples Gram-Hanssen 2010; Reckwitz

2002; Schatzki 2002). This analysis draws upon Elizabeth Shove's (2010) division of practices into three elements of objects (the physical items that are involved or related to a practice), competences (the knowledge, skills, or know-how involved in carrying out a practice), and meanings (the moods, morals, ethics, attitudes, dispositions, goals, and values that are attached to practices).

Practices do not take place in a social vacuum. Normativity, that is, the acceptability of a practice, along with its 'oughtness' or 'rightness' (Schatzki 1996: 101–102), is in part socially situated. The range of acceptable or unacceptable actions, or thoughts or feelings in a given situation, which are shaped by perceived actions, thoughts, feelings, and expectations of relevant others, can be referred to as 'social norms' (Giguère et al. 2016). Social norms can be regarded as the 'grammar of society' in the sense that they are implicit, spontaneously created (i.e., they are not specifically designed) rules that define how the social world operates (Bicchieri 2005: ix). The desire to belong to and the fear of exclusion from reference groups[4] are assumed as key in motivating conformity to social norms (Cialdini & Trost 1998). There is strong empirical backing for social norms, which originates in social conformity experiments during the early to mid-20th century (Cialdini & Goldstein 2004). Social norms can be distinguished as either descriptive (what most others do) or injunctive (what most others think ought to be done) (Cialdini et al. 1990). While these types of norms are posited as conceptually and motivationally distinct (Cialdini et al. 1990), it is often difficult to distinguish between the two, as describing what most other people do 'inevitably introduces *injunction*; that is, "most people do this" becomes "people should do this"' (Burchell et al. 2013: 3).

This analysis combines the use of SPT, drawing on Shove's categorisation of practices into objects, competences, and meanings, with the concept of social norms. The use of SPT draws attention away from the individual, which then allows a focus on what the actions are and how actions are meaning-laden and enabled or constrained by their environment. While the use of social norms enables discussion of individual agencies, they also highlight how perceived group expectations play an important role in shaping sustainability-related practice.

Findings

Sharing spaces

Sharing space is highly effective in lowering domestic greenhouse gas emissions. Through increasing household density, residents lower their per-person energy consumption for lighting and space heating, the latter of which accounts for the largest share of energy use in UK households (BEIS 2020).

However, despite its many benefits, sharing domestic space has also been found to be a challenge. Research into house-sharing has found that common causes of conflict centre around a perceived lack of privacy (Klocker et al. 2012) and the importance of managing the cleanliness of shared spaces (Clark et al. 2017). Indeed, a survey of British house-sharers found that 43% of the respondents reported that at times they stayed in their rooms to avoid interaction with other residents (Avis-Riordan 2018). Sharing space therefore lies at a nexus of being both a social challenge and a pro-environmental opportunity. This section explores how spaces were negotiated in the case study's intentional communities.

At Valley Rise, upon first glance, there was a clear demarcation between shared and private space, with 'private space' defined here as a space in which one is not observed or disturbed by others. Private spaces were residents' bedrooms. All other spaces—a kitchen, two lounges, bathrooms, and a garden—were known as shared spaces: they were available for the use of all residents and occasional guests. Yet, it was found that regarding a *space* as private or shared was reductive (although the terms 'private spaces' and 'shared spaces' are used in this section for expediency). Cohousing communities also have designated 'private' and 'shared' spaces, although, as with the house share community, how spaces are used is more fluid than this simple binary definition.

Within the cohousing communities, residents' homes are self-contained, meaning that in theory residents do not or rarely *need* to venture into shared space. At Bramley Yard, there were numerous communal spaces: two kitchens, meeting spaces, a swimming pool, gardens, and 40 acres of farmland. At Meadow Cohousing there were allotments, garden space, and a common house, which included a kitchen and events space, meeting space, and an office. In cohousing, the greater amounts of space plus the self-contained nature of residents' private space meant that being physically present together was far less common than in the house share community. (Indeed, another study of US cohousing communities found that residents spent approximately only 5% of their time in communal spaces (Williams 2005).

This research also found that communal areas were often empty or perhaps would have only one or two people in them, although residents reported plenty of bumping into one another, which often led to social exchanges. Nevertheless, there were generally not the same challenges of negotiating shared spaces and achieving privacy as there were in Valley Rise. Therefore, this section draws more on examples from Valley Rise, where residents shared more space more often.

Perceptions of the right to privacy within shared spaces could change with a change in the frequency of use of that space. One challenge acknowledged by residents of Valley Rise was that tensions that could arise from misaligned perceptions on the right to use shared spaces for private moments:

> People would get into habits of like, I like to meditate at 7.30 each morning, then they get used to that, and then they think it's their kind of right to meditate at 7.30 in the morning. And then if someone else wants to do it, even just occasionally, then they can feel a bit their noses are a bit out of joint [...] and the other person can feel resentful, because you know, this person uses it 99 percent of the time, and I just wanted to this one... So it's difficult.
>
> *(Harry, Valley Rise, September 2019)*

Here the right to use the shared space is viewed through two different prisms. The first is that of routine: 'I always do this at this time' becomes 'it is my right to do this at this time', which has echoes of a descriptive norm introducing injunctive elements (Burchell et al. 2013). The second prism we apply to the right to shared space is based upon principles of having a 'fair share'. According to this principle, the distribution of privacy within shared space matters, and the fact that one person has the space to themselves regularly gives another person the right to even up this distribution through exercising the same privilege within that space. Both principles are based around temporality, which illustrates that privacy within space has a temporal element: shared spaces can at times be private spaces.

The effect of privacy is also attached to certain practices. In the example above, the need for privacy is attached to the practice of meditation. Similarly, within the cohousing communities, the boundaries of privacy and communality within spaces changed. Meetings were usually open to community members only. Some events were open to the public (such as music nights or open days), and some events, such as those with intimate emotional sharing, might consistently be attended by a smaller group that required privacy within a shared space during those meetings. Privacy was therefore found to not be an attribute of a space, but an attribute linked with temporality, routines, and practices within that space, which, as illustrated, could be contested.

While the right to shared spaces is viewed above through the prism of privacy (a concern of residents that emerged during the fieldwork), it may be more useful to view sharing spaces themselves through the prism of agency, i.e., the ability to choose what action to take, unfettered. To achieve this, an alignment of norms, both in terms of where and when practices 'belong' in each of the communities, was needed.

Part of this alignment of norms in these communities was a matter of scheduling use of space, for example, establishing routines around use of the bathrooms or booking a space in advance. At Valley Rise, this booking usually meant a WhatsApp message to housemates, whereas at Bramley Yard and Meadow Cohousing online or physical calendars were used. These actions required certain competences that are unlikely to be needed in single-family households, including the use of online tools, increased levels of planning, and incorporating the consideration of those who were not attending the event into that planning by informing them of use of space.

Residents at Valley Rise also aligned norms around use of shared space through discussion and agreement.

> We made a rule in the end that was like, it's fine to talk in the kitchen, and it's fine to put the radio on. And if you want, you know, more quiet space, you should be in the living room. So like, you didn't get a right just because you'd started working five hours earlier in the kitchen for other people to not feel welcomed there [...].
>
> *(Harry, Valley Rise, September 2019)*

In the example above, tension was resolved by attaching norms to spaces rather than to practices, although the challenges that residents raised (see the previous section) indicated that this solution was not always deployed. The group did make a great effort to pay attention to their own needs and the needs of others, however, through the use of nonviolent communication (NVC), which is a communication technique that is often used to manage potential conflict. Its basic tenets are about self-empathy—empathy for others and self-expression (CNVC n.d.). It is a well-known method and correlates in intentional communities with high levels of satisfaction about group decision making (Rubin et al. 2019).

> There's someone who has like lots of difficulties sleeping, and so really needed like doors to be closed very quietly after a certain time [...] it was really helpful to like hear why, because that stuff that's like slightly annoying like 'Oh I don't want to close door quietly it's just, you know, it's ten o'clock'. But, really, NVC really allowed us to like hear [...] why [...] it was really important for this person and it wasn't just a little thing was actually [...] [it was] connected to quite a lot of different things in that person's life and [...] that you know helped me change my behavior.
>
> *(Josie, Valley Rise, September 2019)*

NVC enabled those with a request (e.g., that doors be closed quietly at night) to create a framework through which to make that request in a way that encouraged understanding. This practice, consciously adopted by the household, encouraged the sharing and alignment of meanings attached to practices, with positive results.

Sharing Things

Homes are filled with a vast array of things. Imagine for a moment your own home or perhaps the home of your parents and think of not just the furniture, the white goods, the clothes, and electronics, but the games, the books, the papers, the tools, the drawers full of miscellaneous items, and all the things in garages and attics and cupboards. Every one of these things has greenhouse gas emissions attached through the extraction of its raw materials, its manufacture, and its transport to the shop you bought it from or to your front door. Altogether, these emissions are significant, although they are also extremely difficult to measure due to the high level of complexity involved in indirect emissions measurements and the lack of consistency between the extraction, manufacture, and transport of different items.

Yet, to ignore the environmental impact of things (more commonly referred to as *goods*), in a study of intentional communities and environmental impacts would be remiss, given how intentional communities can and often do share things, and the very high impact of this emissions category (found to be 30.6% of average UK household emissions when measuring greenhouse gas emissions from energy and water, food, transport, purchases, and waste (Clark 2021). This section investigates how the case study intentional communities managed the sharing of things.

In all the communities, the sharing of everyday items appeared to be a common practice and took two forms: the borrowing of privately owned items and the sharing of communally or jointly owned items/amenities. During the fieldwork, numerous examples of both occurred, such as Bramley Yard resident Sal sharing a bread maker with her neighbour or Meadow Cohousing resident Peter borrowing his neighbour's television to see whether he wanted to invest in buying his own. Within the study's cohousing communities, most objects were still privately rather than communally owned (with clear delineations between the two). In this sense, cohousing communities maintained the norms of private ownership most of the time.

However, there were certain objects and facilities that were kept in communal areas and available for communal use, including cars, sofas, toilets, solar panels, washing machines, books, DVDs, gardening tools, bike maintenance equipment, and toys. In Valley Rise, where residents shared one house, all items, furniture, and amenities that were not in the bedrooms were communally used, which meant that these items were not viewed as borrowed but rather jointly owned and shared. As such, there were no questions of 'Can I use this?'; rather, sharing many items was an inherent practice and considered part of sharing a house. For example, in the Valley Rise lounge there was a bookshelf that held books belonging to individual residents and gym equipment that belonged to one resident: every item could be used by anybody.

It was clear in all the communities that the sharing of things required various forms of effort, which this next section explores.

As most items were shared sequentially rather than simultaneously, it was important for residents to leave shared items in a condition that was ready for use by the next resident, that is, clean and in its proper storage place (i.e., tidy). Ideas of tidiness and cleanliness were a common topic of conflict within intentional communities; the topic arose several times in interviews with residents at Bramley Yard. For example, more than one person mentioned the conflict over what items should or should not be stored on porches, which were in themselves ambiguous spaces because they were not communal, but were in communal view and adjacent to shared space.

> I knew that somebody would be very unhappy with them being stored there. So, I've been reminded this morning by someone else that they're still there—those potatoes. So there is a pressure from some elements to keep things neat and tidy.
>
> *(Dee, Bramley Yard, October 2018)*

At Valley Rise, the residents went to some effort to create a shared idea of what clean and tidy meant for each room by writing details of what cleaning each room involved on a sheet of paper that was displayed in the kitchen.
Bathrooms

- surfaces: spray + wipe
- mirrors + bath screen: spray + squeegee
- toilet: scrub inside, wipe top, spray + wipe seat
- bath: spray, sponge, remove hair
- put cleaning products away:)

Description of the correct way to clean the bathrooms at Valley Rise
This sheet of paper supplemented the community's shared cleaning schedule. According to some Valley Rise residents, creating this shared understanding of clean and tidy prevented conflict or resentment.

Residents also tried to influence the correct use of shared objects and materials through written notes and signage. This was particularly the case in the cohousing communities. There were a greater number of signs at Bramley Yard Cohousing than Meadow Cohousing, perhaps because Bramley Yard residents took more responsibility for the endeavour together or perhaps because of its being a farm that requires more coordination and more know-how in order to operate efficiently.

Some of this signage had directions with pro-environmental outcomes. For example, a sign at Bramley Yard, which read 'CALABRESE READY TO

EAT' was posted to encourage residents to harvest and eat the crops. Other examples of signs with pro-environmental outcomes included a sign giving instructions on how to switch the heating on and off in a communal area (encouraging heating spaces only when in use) and a sign asking people to keep a door shut to prevent heat loss. Signage was a way in which reminders of desirable practices and their attached injunctive norms could be embedded into the environment; they acted as 'nudges' towards what the author of the sign perceived as appropriate practice.

Indeed, ideas of appropriate practice for the placement, condition, or use of things were not always shared. Notably, in the quote from Dee above, she appears to admit that her potatoes left on the porch were *not* neat or tidy and also knew that at least one person would be very unhappy with them being stored there. While Dee admitted that there is not a shared understanding of how things (in this case, potatoes) should be placed in a space visible to the community, her understanding of communal life encompassed conflict and so she saw this as a norm. She described how, when she was visiting Bramley Yard with the idea of moving there, the community was going through a conflict that centred around whether or not to install a district heating system:

> We were also visiting across a quite difficult time. So... probably our second, maybe third visit, everything had blown up about the boiler being put in [...] we heard loads of different perspectives from different people [...] And it was great! Because we had this really honest idea of how hard it can be. So [...] we always knew [...] there would be really annoying aspects.
>
> [...]
>
> the way I see it, is, people, everybody's, everybody comes from a really positive place. Everybody here is doing, I think, everything with the right reasons [in mind] [...].
>
> *(Dee, Bramley Yard, October 2018)*

Dee's acceptance that conflict was a part of communal life was grounded in the belief that people all 'come from a positive place', as she put it. In this sense, she believed there was a shared value of good intentions, if not a shared understanding. Her perspective and experiences demonstrate that, while there is always the potential for conflict, it may not always be something that has to be solved. A way of sharing may be disagreeing about how things are shared but doing it anyway, accepting that disagreement is part of communal life.

In the two larger communities, Bramley Yard and Meadow Cohousing, systems were in place for the community-wide sharing of some things, such as tools, and indeed, there were gripes and frustration about how things were shared. For example, during a Bramley Yard community meeting:

> There was a lot of joking and groans when Andy brought up the matter of a system for keeping track of tools right at the end of the meeting.
>
> *(Fieldwork journal, October 2018)*

Similarly, during a sustainability workshop at Meadow Cohousing, where residents were discussing how to increase sharing practices, residents spoke about not being able to find communal tools and agreed that they:

> have to buy in to sharing and put stuff away properly, maintain it and advertise that it's for use. Have a system in place.
>
> *(Fieldwork journal, November 2018)*

In both cohousing communities, some residents suggested that a system was needed to be able to successfully share tools, but noted that agreements (and adherence to those agreements) about the correct placement and condition of things could be difficult to arrange and maintain. Notably, during the aforementioned Meadow Cohousing workshop, during a group discussion on how sharing practice could be improved, the efforts required to improve sharing were acknowledged during discussion:

- We should make efforts to not just dump stuff, but to have expectations of the things we share;
- Think of easy things to share, such as books;
- Part of sharing more is having knowledge about who to go to for something;
- We should have a better system in place for borrowing the communal tools, though with all sharing systems, they shouldn't be too complex. We should think about what systems are realistic given our energy levels;
- The bike sheds are a collective responsibility, but sometimes this means that nobody takes responsibility. Unless people really take responsibility for a bike pool, there shouldn't be one; and
- There are different modes of sharing: private possession and borrowing and things which belong to everyone. We should think about which items fit best into which category. (Fieldwork journal, November 2018)

Altogether, this section demonstrates that some of the labour that goes into the sharing of things involves the development of systems and norms, competences, knowledge, and emotional and relational work.

Another element that this workshop discussion highlighted was the intentionality of sharing at Meadow Cohousing, which was also observed in the other communities. Negotiating these intentional shared practices prompted actions to become salient, which, as some studies have shown, makes actions more likely to be impacted by norms (Cialdini et al. 1990), including pro-environmental norms. This intentionality around sharing was framed by certain meanings attached to sharing, such as greater efficiency, 'You don't need one of everything, there just needs to be enough of things spread amongst the community' (Tom, Bramley Yard October 2018) and environmental sustainability ('lesser impact on available resources, both human and non-human' (Peter, Meadow Cohousing December 2019). Indeed, the quantitative data arising from the fieldwork indicated that the average household at Bramley Yard and Meadow Cohousing spent less money on clothes and shoes, small household items, electronics, and transport-related equipment than the average UK household (see Clark 2021: 183, Figure 6–13). (Note that Valley Rise did not complete expenditure surveys, so no data on the environmental impact of the community's things was collected.)

These lower-than-average figures were attributable to meanings of anti-consumerism held by many residents. Borrowing and sharing practices, enabled by communal life, also carried meanings of anti-consumerism and sharing. Although they were sometimes difficult and could require additional labour, they were still regarded positively. Take, for example, this comment by Judy at Bramley Yard, who shared that she found the communal carpooling to be 'a joyous community endeavour'.

Discussion

The sharing of spaces and things currently most often exists either at a micro level within households, the majority of which are family units, or at a macro level, consisting of public spaces and things that are managed by local or national governments. (The latter include pavements and other paved spaces, public transport, parks, etc., or privately owned spaces and things managed by companies or perhaps even powerful individuals (e.g., offices, shops, restaurants, etc.). However, sharing and collective management of spaces and things does not tend to exist at the meso level of neighbourhoods, and while it does exist in non-related house shares, it can often prove challenging (Clark et al. 2017; Klocker et al. 2012) and is generally seen as a transitional life stage before progression towards a household centred around a nuclear family.

Intentional communities offer us scalable models and methods for sharing at the meso level and/or between non-related people that enable us to reconceptualise what home can be and mean. This discussion begins by highlighting key elements that constitute this reconceptualisation of home, namely, the implementation of salient rules and systems, the communalising of domesticity, and the replacement of the nuclear family as the epitome of 'home'. The discussion then goes on to consider how models and methods can be adapted for those outside of intentional communities, thus enabling the scaling out of sharing and collective management.

One typical practice of intentional communities is co-created documentation of shared values and goals (Meltzer 2005). This practice in domestic space is unusual; more directive instructions (i.e., rules) on how to act in households tend to be applied only to children (Holloway & Valentine 2001) or those employed by the household, such as au pairs (Cox & Narula 2003). Such documentation of shared values and goals is created to protect certain doings and meanings in the home that ensure the comfort of the rule-makers (Cox & Narula 2003; Holloway & Valentine 2001).

There is variation on how directive the documentation co-created by communities is. Some documents only contain certain values that are embodied by the community, whereas others include specific rules, e.g., relating to pets or to how much time people are expected to commit to communal work. Yet, whether directive or not, to some extent they appear to serve a purpose similar to rules in that they make embodied meanings explicit (Cahill 2000).

The bringing of rule-making practice into the domestic sphere is one that is reasonably unique to intentional communities and may be at odds with meanings of home, in which practices between adults are generally based on implicit norms and are perhaps negotiated informally, but not usually detailed explicitly, and certainly not written down during a meeting. Indeed, during fieldwork, some participants expressed unease with the notion of rules at home (in relation to rules around sustainability), although, in reference to their house agreement, Harry from Valley Rise said that 'rules are freeing' and provided the example that with an agreed-upon cleaning rota, nobody needed to feel guilty that they were not taking part in cleaning when it was not their turn to do so (Harry, Valley Rise October 2019). This view, at odds with a more typical perception that rules constrain, demonstrates how the meanings around rules can be reconceptualised to be linked with relaxation and emotional ease by removing uncertainty around group norms. The co-creation of rules makes it more likely that practices will reflect (shared) norms (Cialdini et al. 1990), which may result in more pro-environmental outcomes if enough members of the group share environmental values.

Intentional communities also tend to make use of practices more typically seen in organisations such as workplaces, namely, meetings (often accompanied by agendas, notes and actions) and, in the case of larger communities (e.g., cohousing), governing systems to manage coordination of activities. Bringing workplace practices into the domestic realm opens up new possibilities for coordination and collaboration beyond the nuclear family, enabling or at least easing the achievement of shared goals, whether those goals are having regular group meals, creating a community garden, or conducting political canvassing. Bringing these practices to homes makes homes the site of collaborative empowerment and extends what is possible in the domestic sphere.

Whereas most of domestic life is now intensely private, intentional communities draw domestic practices (such as cooking, cleaning, gardening, and child-rearing) out from the isolation of the nuclear family into the presence of other unrelated people. As such, Helen Jarvis (2015: 98) writes that 'mundane tasks [...] are transformed by the sociality of work as a ritual that is performed and endowed with a sense of occasion'. Therefore, domestic practices have the potential to become events with meanings of sociability and conviviality attached. Shared domestic practice also creates opportunities for the sharing of competences, whether in the form of learning new skills or one person helping another (e.g., fixing their bicycle), which may result in pro-environmental outcomes if individuals carry pro-environmental competences.

The concept of home is closely linked with the nuclear family. It is typically envisioned as a heterosexual couple with children, an image which persists despite widespread acknowledgement that families may take very different forms and that the majority of households in the UK contain less than three people (ONS 2021). The persistence of this image also ignores the fact that for those who do live in this typified nuclear household, it is just one phase in a lifecycle which includes other types of households, rather than an 'ultimate destination' in family composition.

The nuclear family model both assumes a self-sufficiency that places a lot of social and material strain on the nuclear family, while also regarding the needs of households that do not fit that mold as atypical or secondary. Intentional communities move this focus of home from the nuclear family to either other forms of family (e.g., households of unrelated adults) or to the neighbourhood level. This adaptation in meaning of what 'family' or 'home' look like acknowledges more effectively the variations in household composition and the benefit of mutual support between households of unrelated people.

While intentional communities remain a niche form of living, these reconceptualisations of the meanings and competences around home and the sharing of spaces and things can be scaled out by being incorporated by households outside of intentional communities. This final section offers some reflections on this issue.

Firstly, a reflection on the sharing of spaces and things that can be taken from these communities is an acknowledgement of the labour attached to sharing and the understanding that sharing is a competency that involves knowledge, understanding, embodied know-how, and also requires emotional and relational work. On these two latter points, there can be significant negotiation required between sharers to deal with or live with conflict on the arrangements and correct uses of what is shared. Recognition of this work is key to engaging in successful long-term sharing. Without this, systems are likely to fall apart, and people are likely to eventually revert back to private ownership once they are able to do so. As such, the implementation of novel systems of sharing should be accompanied by support for those sharing in order to develop the competences and capacities needed to do so.

Even given these requirements of various forms of labour, cohousing communities and other forms of intentional community are sites of inspiration for sharing. They can show people outside these communities what is possible, such as local-level carpooling and bike-sharing, the sharing of certain objects such as toys, tools, and gardening equipment, use of communal space, and the coordinated growing of food. Not only can they demonstrate what is possible, they can also show how systems of sharing can work and can share know-how about where snags and difficulties in such systems tend to occur.

If a neighbourhood is interested in holding a shared meal, for example, they could copy and adapt the systems that cohousing communities have in place (involving elements such as sign-up sheets, cooking rotas, how to note allergies and other food needs, etc.). Typically, visitors to such communities are shown how those communities 'do things'. What if those communities showed how they do things, but with a framing of 'and this is how you could do it, too, even if you don't live here'? Should this be an attractive notion to intentional community residents, there is much about the systems of sharing they have in place that can be translated to housing located outside their communities.

A key element to encourage more forms of sharing is to shift the meanings around sharing. This is something that the intentional communities involved in this research appeared to do. Rather than sharing being conceived as something people did out of economic necessity, attached to which are meanings of being poor and lacking the freedom to choose, intentional community residents could attach more positive meanings such as reduced environmental impacts and greater efficiency.

Sharing can also be recognised as an enabler of social connection and (as mentioned above) a skill worthy of development. Positive attributes attached to sharing are also likely to impact how sharing is carried out. For example, if people regard sharing as a way to connect, they may be more likely to donate what they regard as 'nice' things to the community as opposed to simply discarding what they no longer want into communal ownership. They may also be more likely to take care of shared objects if they believe that sharing is a skill deserving of their time and attention. Encouraging such meanings and narratives around sharing may act to shift its status towards being aspirational. Arguably, cohousing communities and other forms of intentional community have already promoted this different interpretation of sharing, although governments and organisations can do more to elevate what is shared and also offer more meso-level, place-based spaces and amenities for local communities to share and take pride in.

Intentional communities demonstrate that things can be done differently; they open up the potential for unrelated people to share in a more familial way and for meso-level sharing between neighbourhoods. In particular, the encouragement of collective ownership and management of some spaces and things between neighbourhoods could shift the perception of what being a citizen means. Such encouragement could also offer greater citizen engagement and empowerment, place-based and place-sensitive collective management of resources, stronger connections between people within a place, a more resilient community, and lower environmental impacts. A collective management model could achieve this with little or no intervention from government, non-profits, or private companies.

Conclusion

This chapter explores how spaces and things were shared in three intentional communities. It argues that the lessons learned from these communities could be scaled out by applying them outside of intentional community settings. Given the large and growing populations of cities (World Bank Group n.d.), issues around lack of affordable housing and the urgent need for both climate change mitigation and adaptation, it is likely that more sharing of resources will happen, either out of necessity or the desire for greater efficiency. Intentional communities can help us revitalise our narratives and competences around sharing, and posit it as an important social connection that offers personal development, and results in simply being kinder to both human and non-human beings.

Notes

1 When measuring greenhouse gas emissions from energy and water, food, transport, goods, and waste.
2 Please note that names of the intentional communities are pseudonyms.
3 Cohousing is defined here as a group of self-contained homes plus shared communal space, most often in the form of a common house and a self-managed community which intentionally shares resources, experiences, and social time.
4 Reference groups can be defined as friends, family, or relevant others with whom an individual compares and evaluates their own practices and meanings (Koger & Du Nann Winter 2010).

References

Avis-Riordan, K. (2018, 13 April). This is what makes the perfect flatmate, according to Brits. *House Beautiful*. www.housebeautiful.com/uk/lifestyle/a19753408/whatmakes-perfect-flatmate-uk/

Bamford, G. (2001). Bringing us home: Cohousing and the environmental possibilities of reuniting people with neighbourhoods. In R. A. Hindmarsh (Ed.), *Situating the environment at The University of Queensland* (pp. 36–43). The University of Queensland. https://espace.library.uq.edu.au/view/UQ:13658

BEIS (Department for Business, Energy & Industrial Strategy). (2020). *ECUK: Consumption data tables. Chart 1, consumption by sector; 1970, 1990, 2019* [Microsoft Excel spreadsheet]. www.gov.uk/government/statistics/energy-consumption-in-the-uk

BEIS (Department for Business, Energy & Industrial Strategy). (2023). *2021 UK greenhouse gas emissions, final figures*. https://assets.publishing.service.gov.uk/government/uploads/system/uploads/attachment_data/file/1134664/greenhouse-gas-emissions-statistical-release-2021.pdf

Bicchieri, C. (2005). *The grammar of society: The nature and dynamics of social norms*. Cambridge University Press.

Burchell, K., Rettie, R., & Patel, K. (2013). Marketing social norms: Social marketing and the 'social norm approach. *Journal of Consumer Behaviour*, 12(1), 1–9.

Cahill, C. (2000). Street literacy: Urban teenagers' strategies for negotiating their neighbourhood. *Journal of Youth Studies*, 3(3), 251–277.

Centre for Nonviolent Communication. (n.d.). *What is nonviolent communication?* www.cnvc.org/learn-nvc/what-is-nvc

Cialdini, R. B., & Goldstein, N. J. (2004). Social influence: Compliance and conformity. *Annual Review of Psychology*, 55, 591–621.

Cialdini, R. B., & Trost, M. R. (1998). Social influence: Social norms, conformity and compliance. In D. T. Gilbert, S. T. Fiske, & G. Lindzey (Eds.), *The handbook of social psychology* (pp. 151–192). McGraw-Hill.

Cialdini, R. B., Reno, R. R., & Kallgren, C. A. (1990). A focus theory of normative conduct: Recycling the concept of norms to reduce littering in public places. *Journal of Personality and Social Psychology*, 58(6), 1015–1026.

Clark, P. I. (2021). *Practices of shared living: Exploring environmental sustainability in UK cohousing, community living, and coliving*. [Doctoral dissertation, University of Westminster].

Clark, V., Tuffin, K., Frewin, K., & Bowker, N. (2017). Shared housing among young adults: Avoiding complications in domestic relationships. *Journal of Youth Studies*, 20(9), 1191–1207. https://doi.org/10.1080/13676261.2017.1316834

Climate Change Committee. (2020). *The sixth carbon budget: The UK's path to net zero*. www.theccc.org.uk/wp-content/uploads/2020/12/The-Sixth-Carbon-Budget-The-UKs-path-to-Net-Zero.pdf

Cox, R., & Narula, R. (2003). Playing happy families: Rules and relationships in au pair employing households in London, England. *Gender, Place and Culture*, 10(4), 333–344.

Daly, M. (2017). Quantifying the environmental impact of ecovillages and co-housing communities: A systematic literature review. *Local Environment*, 22(11), 1358–1377.

Department for Business, Energy & Industrial Strategy. (2020). *ECUK: Consumption data tables. Chart 1, consumption by sector; 1970, 1990, 2019* [Microsoft Excel spreadsheet]. www.gov.uk/government/statistics/energy-consumption-in-the-uk

Department for Business, Energy & Industrial Strategy. (2023). *2021 UK greenhouse gas emissions, final figures*. https://assets.publishing.service.gov.uk/government/uploads/system/uploads/attachment_data/file/1134664/greenhouse-gas-emissions-statistical-release-2021.pdf

Foundation for Intentional Communities. (n.d.). *About FIC*. www.ic.org/foundation-for-intentional-community/

Giguère, B., Sirois, F. M., & Vaswani, M. (2016). Delaying things and feeling bad about it? A norm-based approach to procrastination. In F. M. Sirois, & T. A. Pychyl (Eds.), *Procrastination, health, and well-being* (pp. 189–212). Academic Press.

Gram-Hanssen, K. (2011). Understanding change and continuity in residential energy consumption. *Journal of Consumer Culture*, 11(1), 61–78.

Gram-Hanssen, K. (2010, November). Introducing and developing practice theory: Towards a better understanding of household energy consumption. In *Proceedings of the sustaining everyday life conference: April 22–24, 2009*, No. 038, pp. 45–57. Campus Norrköping.

Holloway, S. L., & Valentine, G. (2001). Children at home in the wired world: Reshaping and rethinking home in urban geography. *Urban Geography*, 22(6), 562–583.

House of Commons Business, Energy and Industrial Strategy Committee. (2022). *Decarbonising heat in homes* (HC 1038). Parliament, UK. https://committees.parliament.uk/publications/8742/documents/88647/default/

Ivanova, D., & Büchs, M. (2020). Household sharing for carbon and energy reductions: The case of EU countries. *Energies*, 13(8), 1909.

Jarvis, H. (2015). Towards a deeper understanding of the social architecture of cohousing: Evidence from the UK, USA and Australia. *Urban Research & Practice*, 8(1), 93–105.

Klocker, N., Gibson, C., & Borger, E. (2012). Living together but apart: Material geographies of everyday sustainability in extended family households. *Environment and Planning A*, 44(9), 2240–2259.

Koger, S., & Du Nann Winter, D. (2010). *The Psychology of environmental problems: Psychology for sustainability*. 3rd ed. Psychology Press.

McCamant, K., & Durrett, C. (2011). *Creating cohousing: Building sustainable communities*. New Society.

Meltzer, G. (2005). *Sustainable community: Learning from the cohousing model*. Trafford Publishing.

Office for National Statistics. (2021). *Household and resident characteristics, England and Wales: Census 2021*. www.ons.gov.uk/peoplepopulationandcommunity/householdcharacteristics/homeinternetandsocialmediausage/bulletins/householdandresidentcharacteristicsenglandandwales/census2021

Reckwitz, A. (2002). Toward a theory of social practices: A development in culturalist theorizing. *European Journal of Social Theory*, 5(2), 243–263. https://doi.org/10.1177/13684310222225432

Rubin, Z., Ludwig, Y. & Willis, D. (2019). Conflict resolution and satisfaction in today's intentional communities. *Communities*, (184), 57–60.

Schatzki, T. R. (2002). *The site of the social: A philosophical account of the constitution of social life and change*. Pennsylvania State University Press.

Schatzki, T. R. (1996). *Social practices: A Wittgensteinian approach to human activity and the social*. Cambridge University Press.

Shove, E. (2010). Beyond the ABC: climate change policy and theories of social change. *Environment and Planning A*, 42(6), 1273–1285.

Underwood, A., & Zahran, S. (2015). The carbon implications of declining household scale economies. *Ecological Economics*, 116, 182–190.

Williams, J. (2005). Designing neighbourhoods for social interaction: The case of cohousing. *Journal of Urban Design*, 10(2), 195–227. www.tandfonline.com/doi/pdf/10.1080/13574800500086998

World Bank Group. (n.d.). *Urban development: Overview*. www.worldbank.org/en/topic/urbandevelopment/overview

8

COMMUNITARIAN 'INHABITING'

Empathic forms and conflict resolution spaces

Sylvia Amar

Introduction

Context

In critical architectural history, the concept of Inhabiting refers to the interaction between built space and social practices that reflect ways of life, social structures, and cultural identities (Leroux 2008). Inhabiting goes beyond the mere occupation of places to include symbolic meanings and dynamics of spatial appropriation (Paquot 2005; Paquot, Lussault, & Younès 2007). More than that, Inhabiting[1] is a way of being authentically in the world, an integral experience in which building and living unite in the creation of places that allow humans, as individuals in communities, to fully exist (Heidegger, Préau, & Beaufret 1973). Inhabiting is living in a project that is simultaneously social and spatial and supported by a set of both structuring and unifying values and rituals.

Although not very visible to the general public, the communitarian world occupies a central place in this reflection. Most often located in rural areas, it is composed of entities that define themselves, sometimes doubly, when such communities explicitly identify both as intentional communities and as eco-villages. This is the case with Findhorn Ecovillage (www.findhorn.org), Dancing Rabbit Ecovillage (www.dancingrabbit.org), and Sieben Linden Ecovillage (www.siebenlinden.org). These are semantic evolutions that denote a continuous movement. At least since the 19th century, this global movement has unfolded in regular waves tinged with ideological motivations (socialism, anarchism, ecologism) depending on different historical and geographical contexts (Creagh 2009).

DOI: 10.4324/9781003769064-9

From a chronological point of view, the term 'intentional communities' was used up until the mid-1990s. The emergence of the concept of eco-village is credited to the Danish Gaia Trust Foundation which, after commissioning a study by Robert and Diane Gilman, started to support the creation of several communities by providing funds to help establish them and develop their economic activities. These two American researchers and editors of the magazine *In Context* launched the movement that built the Global Ecovillage Network (GEN) during the famous Findhorn Conference held in 1995 in Scotland in the eco-village of the same name. Some intentional communities already in existence gradually adopted this new designation, while new initiatives used it immediately.

Intentional communities and/or eco-villages present themselves as alternatives to the dominant urban way of life often marked by individualism and high-carbon lifestyles. While diverse and shaped by their geographical contexts, these communitarian experiences share a common feature: the development of a social project that is intentionally embodied—both physically and symbolically—in a spatial one. This socio-spatial project engages multiple, overlapping scales of Inhabiting, ranging from the built environment to the territory, from daily practices to cultural values, and from relational dynamics to emotional experience. While this approach may echo in part the multi-scale thinking of Christopher Alexander's *Pattern Language* (Alexander 1977), the emphasis here is less on systemic patterning and more on the lived, socio-emotional layers of ecological dwelling.

I argue in this chapter that empathy is central to the physical and social design of contemporary eco-villages. I discuss two dimensions of empathy: one linked to design and the other linked to social relations. According to Van der Ryn:

> Empathic design implies thinking ahead, integrating probable future risks such as oceans rising, temperatures rising, soils declining in fertility, chemical pollution of water. Empathic design should consider both the precautionary principle as well as the law of unintended consequences.
> *(Van der Ryn 2013: 8)*

Socially, empathy constitutes an active mechanism for understanding other people's needs and perspectives, thereby fostering constructive interactions and greater social cohesion (Beltrán Gutiérrez 2024; Brown et al. 2019; Berardi et al. 2020). Because it is able to connect with the experiences of others while maintaining an awareness of one's own subjectivity, empathy acts as a catalyst for relationships based on reciprocity and mutual recognition.

It can be said that eco-villages implement highly specific local systems capable of introducing positive effects on the planetary conditions of future generations (Jonas 1979). They are considered here as alternative places, but also as ecosystems which, in order to be built and endure, must face a diversity of relational situations, including interpersonal tensions, collective decision making, and interactions with surrounding communities. If their modes of governance—most often participatory and/or shared—establish a horizontal and fluid framework, the resolution of internal and external conflict requires attention that is more empathic than organisational.

The implementation of the concept of empathy establishes a coherent link between the creation of socio-spatial modes of Inhabiting (physical, ethical, sensitive) and the awareness of the impacts of human activities on the environment. Communitarian spaces are living, evolving systems that are constantly adapting. Architecture—considered here as a set of situated practices shaping space, use, and social interaction—thus becomes a privileged formal and conceptual language which, without resorting to gesture or style (as Van der Ryn advocates), opens up perspectives on the introduction of changes necessary for the survival of the system itself. Given the global ecological context, it would seem logical for the field of architecture to value and draw inspiration from this empathic mode of functioning in its practices and teachings.

Methodological and analytical framework

This chapter explores the subject of communitarian Inhabiting in Europe and the United States (US) through a selection of four cases: La Cité Écologique, established in Canada in 1983 and in the US in 2003, Torri Superiore, established in Italy in 1997, Hallingelille, established in Denmark in 2005, and Le Hameau des Buis, established in France in 2011. These communities are identified according to three shared criteria: the sustainability of their experience, their commitment to ecological architecture, and their openness to global issues. The research presented here is part of a longer study (Amar 2020) that includes four other eco-villages which are not part of this analysis.[2]

The data collection tools of this research include interviews, archives, and field observations that are analysed using comparative tables. Although the sample reflects a diversity of eco-village experiences, its representativeness remains relative; the appearance and disappearance of one in ten such experiences is a major difficulty for research in this field (Christian 2015). The approach favours qualitative over quantitative results, drawing from communitarian modes of everyday, practice-based action.

The theoretical framework of this chapter is based on the concept of 'Place Work Folk' developed by Patrick Geddes, a Scottish biologist, sociologist, and pioneer of modern urban planning, and on its later extensions within American bioregionalism, which in turn inspired European territorialism. In the US, the bioregionalist movement emerged in American countercultural circles as early as the 1960s. Combining scientific and cultural approaches rooted in biogeography and social ecology respectively, bioregionalism advocates a return to basic principles such as 'The world is made of places' (Gary Snyder, poet, and environmental thinker) or 'Think little' (Wendell Berry, writer and agrarian essayist). It developed in an experimental and concrete way in certain communities of the 1970s, such as the New Alchemy Institute in Massachusetts (McGinnis 1998). Territorialism can be understood as a European transposition of bioregionalism: it proposes a return to territory as a way to problematise spatial planning, that is, the relationships between settlement patterns, production systems, and local resources within a global ecological context (Magnaghi & Bonneau 2014; Magnaghi 2017).

Often in motion, the communitarian world has the particularity to offer a field of architectural experiences that synthesise human needs and ecological necessities in order to produce modes of Inhabiting that are coherent with a bioregional awareness of space (McGinnis 1998; Lockyer & Veteto 2013; Magnaghi 2017). One of the main self-declared objectives of intentional communities is to regenerate natural and social environments (Joubert et al. 2015).

In the 1990s, several European countries saw the emergence of architectural productions, understood here as alternative forms of ecological construction and self-organised housing. These developments signalled a return to more traditional modes of organisation and building practices while simultaneously demonstrating increased resilience and energy efficiency at the scale of inhabited environments. Intentional communities from this period can be understood as heirs to earlier communitarian movements. While their spatial and social arrangements are in many respects comparable to those of previous experiments, they simultaneously engage with contemporary concerns related to the climate emergency as well as with environmental, identity-based, and ideological struggles inherited from the 1960s–1970s countercultural movements (Petitfils 2011; Creagh 2009; Amar 2020).

Today, an interlocking of architectural scales, understood as the articulation between inhabited spaces, territorial contexts, and global ecological horizons, expresses the desire of many currently active eco-villages to commit to more sustainable ways of life. Intentional communities are increasingly seen as providing solutions that foster more resilient life systems. This observation reinforces the need to turn our attention toward concrete experiences that may be considered 'real potentialities of humanity' (Wright, Farnea, & Peschanski 2017) in architectural research, and by extension in professional practice and architectural education.

Overall, the approach developed in this chapter is informed by the work of Van der Ryn (1935–2024), particularly his contributions to ecological and empathic design (Van der Ryn & Cowan 2010). Building on broader research about eco-villages developed in a doctoral dissertation (Amar 2020), this chapter advances two main lines of argument. Firstly, the alternative functioning of eco-villages is analysed as an act of rupture with dominant ways of living and as an empathic reinvention of communitarian life (ways of relating to others and to the environment). Secondly, eco-villages are examined as part of a global network that enables them to engage with the world's totality (a shared understanding of global interdependence) through network-based forms of organisation. This mode of organisation contributes to the emergence of new sensory geographies (new ways of experiencing and organising space) consistent with the empathic dimension of contemporary forms of communitarian inhabiting. The following section develops these concepts further and relates them to the concrete functioning of eco-villages.

An act of rupture and empathic reinvention

One of the primary functions of architecture is to respond to the needs of Inhabiting through the design and construction of forms that are in alignment with those very needs. Because the communitarian world once acted without prior conceptual framing of its actions, it has intuitively taken hold of this empathic vocation and turned it into a language in its own right: the physical forms specific to eco-village buildings maintain a strong dialectic with the fundamental needs of the project they host, support, and help to flourish.

Conversely, architecture devoid of empathy may result in buildings designed solely for economic or technical efficiency with no attention paid to actual uses, to the emotions of inhabitants, or to their well-being. Certain building typologies (large standardised complexes and commercial spaces) can oppress rather than welcome; they can fragment and isolate rather than foster community bonds. Communitarian architectural productions thus provide a constructive and spatial lens through which to read this dichotomy. These include small-scale construction, which is often self-built or collectively renovated, that uses local materials and favours vernacular forms adapted to local climates, materials, and building traditions that integrate harmoniously with their surroundings. Before addressing its formal aspects, this form of architecture deserves to be clarified and described in its socio-economic, cultural, and empathic dimensions.

Low economies in search of connection

The involvement of communities in an architecture of Inhabiting is first and foremost an act of rupture and reinvention. Choosing to live in an eco-village represents an alternative decision to leave the mainstream of society, i.e., a physical withdrawal from the standardised and dominant urban lifestyle of our time. Yet, if people leave their jobs and seek reinvention, how do eco-village inhabitants live, and on what?

For eco-villages, economy and ecology represent two spheres that must be linked on a daily basis. The empathic approach, as it is understood here, requires eco-village inhabitants to orient their choices according to how they intend to respond to their responsibility toward future generations (Jonas 1979) and based on their awareness of the effects of human activities on the biosphere and its ecosystems.

Eco-responsible craftsmanship plays an important role in the everyday economic and social organisation of these communities, particularly at La Cité Écologique, a Canadian eco-village where specific spaces are dedicated to creative and productive activities, including clothing recycling and the making of ethical objects such as handcrafted household items or locally designed goods. At the two other eco-villages under study, Torri Superiore and Le Hameau des Buis, micro-scale food production lies at the heart of their communitarian activities in collective kitchens or shared facilities that are used for producing foodstuffs, including olive oil, jams, local wild plants, vegetables, cheeses, and bread. Open-field agriculture is also highly developed at these three sites, which illustrates that breaking from conventional society in favour of a rural way of life is a lived and daily reality.

These eco-villages also host the headquarters and offices of eco-responsible companies working in the fields of energy or recycling (Cités Écologiques in Ham-Nord and Colebrook), as well as groups of freelance professionals. There have been osteopaths in Hallingelille since 2018, which indicates the presence of more conventional forms of work. Less visible but valued by eco-village residents, remote work is also commonly practiced several days a week. This choice is based on reducing work expenses and the environmental impact of commuting while taking advantage of the comfort offered by ecological and/or bioclimatic housing. This comfort was planned at the earliest stages of the founding of Hallingelille and Le Hameau des Buis eco-villages.

It is also worth noting that service and educational activities occupy an important place in the economy of these communities. On the one hand, they are oriented toward technical training related to rural and communitarian activities, including permaculture and training for community development. On the other hand, they focus on training in 'tools of empathy' such as nonviolent communication, environmental education, and conflict resolution. When observing these various activities from the perspective of the forms of Inhabiting that are of interest here, it appears in the eco-villages we studied that subsistence activities (food production) and entrepreneurial activities that create links with the immediate surroundings and the territory remain consistent with the objectives of the communitarian projects. From a spatial perspective, Inhabiting thus integrates the surrounding natural environment (outdoor spaces dedicated to agriculture), technical spaces (the construction or restoration of multifunctional buildings dedicated to craft, training, and educational activities), as well as private spaces (housing) that might also accommodate independent professional activities (Figure 8.1).

The economic structure of these eco-villages informs us about their organisational systems, which are comprised of all collective or individual activities and incomes necessary for a collective subsistence. The ecological dimension in the inhabitants' minimising their needs is reinforced by a set of values (i.e., the pursuit of food autonomy and energy self-sufficiency) that is articulated through the structuring of the project and its modes of governance (the status of the land, ownership of the buildings, decision making, and participation).

Except in specific cases, the eco-villages cited here are focused on providing for themselves. When they market any surplus at very small and local scales to outside populations, they contribute to local exchanges and create human connections with people who reside outside their communities. Thus, when La Cité Écologique takes part in a farmers' basket distribution network, it is because a sufficient level of food production for its population has been reached. This commitment also contributes to local integration into neighbourhood life, which is always carefully maintained. Similarly, at Torri Superiore, if the hospitality activity invites visitors to enter the intimacy of its spaces and uses through the micro-sale of a few local specialties, it is

FIGURE 8.1 La Cité Écologique eco-village, Canada

indeed to promote a porosity between its rootedness in a specific place and its connections with the surrounding environment. One might speak here of a level of empathy through exchange.

The diversity of examples described across different sectors of activity demonstrates that beyond contextual differences between the communitarian projects, there exists a shared micro-territorial dimension. This dimension refers to the scale at which everyday practices, spatial organisation, and the management of local resources are articulated within a clearly situated territory. It establishes an architectural language expressed through spatial scales, material choices, and modes of territorial integration, which highlights how communitarian systems can function as places where 'an economy and an ecology of life' seek to be reconciled in relation to a territory.

The eco-villages discussed here do not intend to escape the reality of the capitalist market, but they do introduce an important nuance by drawing inspiration from models of voluntary simplicity and degrowth. The economic structures of communitarian Inhabiting would therefore be empathic in that they embody the necessary consideration of the flows that pass through them, i.e., monetary and commercial flows on the one hand and systemic and living flows on the other. The way these economic structures are built and deployed reflects an awareness of the dynamic that regulates the necessary interactions between all living systems present within a territory. However, beyond shared features, each communitarian initiative stands on its own. Even when similar activities are present across different eco-villages, they are embedded in particular local contexts that shape how these practices are implemented and experienced.

Shared governance and the production of socio-spatial space

While observation of the sites confirms that living in an eco-village requires breaking away from conventional urban lifestyles, reinventing locally grounded economies with a rural orientation and managing them according to ecological values cannot operate independently without an essential binding factor: the modes of governance practiced.

This description of the governance modes in the eco-villages studied is deliberately synthetic in order to favour an analytical approach. Although the respective modes of functioning vary among communities, the formation of reflection circles composed of inhabitants involved in thematic committees or project-based groups and decision making by consensus constitute shared reference points.

In this study, decision making by consensus is the most widespread form of decision making. This governing mode abides by the rule that requires between 80% to 100% voter approval for proposals to be passed. Because such a large majority is required, a proposal is very rarely rejected outright.

Instead, it returns to a working group for review and possibly reformulation and is then presented again before the collective. In general, a leadership body, a council composed of individuals with different statuses (founders, residents, external members), meets once or twice a year to define broad directions for the community.

Eco-villages also establish regular moments in their daily schedules that are dedicated to residents' group discussions. Several intentional communities, including Torri Superiore, Hallingelille, Le Hameau des Buis, and La Cité Écologique, have developed hybrid and participative governance systems that often combine regular assemblies, thematic working groups, and decision making by consensus. These structures allow for adaptable, skill-based role distribution and intergenerational transmission of practical knowledge and values that range from everyday rules to ecological principles and shared ways of living that together constitute the foundations of empathy.

But the most tangible aspect of the forms of communitarian Inhabiting is embodied in the integration of various activities into fluid and multifunctional socio-spatial forms that are described as *soft*, i.e., flexible, adaptable, and affective, and also low-intensity—evolving spaces that evoke empathy in the environments that host them. Thus, the layout of exterior spaces often resembles the art of gardens (agriculture, agroforestry, animal husbandry), while the buildings constructed or renovated are small in scale and follow traditional local building techniques (housing, schools, common houses), all while accommodating multiple uses (the fields can become spaces for learning and the communal dining halls become classrooms or welcoming spaces for visitors) (Figure 8.2).

This method of producing space (construction, occupation, integration) is so widespread in the context of eco-villages that it can be considered constitutive of the act of rupture and reinvention previously discussed. It operates on the same level as the choice of rural life grounded in artisanal and degrowth-oriented economies, *soft economies* that emerge directly from these situated ways of producing and inhabiting space. Moreover, in addition to the web of daily socio-spatial interactions encouraged by the multifunctionality of spaces, the small number of residents in eco-villages highlights how they are truly micro-living units. Nevertheless, their organisation is no less complex: it requires both individual and collective commitment to cement and articulate the broader ecological ambition. Even more, each community relies on another meta-requirement: reconciling an open social system with a relatively closed spatial experience, if not constrained in its ambitions of self-sufficiency.

This quest for balance reflects the tensions and limitations inherent in the very process of consensus as the cornerstone of shared governance systems and also of the communitarian project and its empathic dimension. Still, a quick look back reassures us about the fragility of shared governance

FIGURE 8.2 Torri Superiore eco-village, Italy

systems and shows that while these systems are *soft*, they are nonetheless robust. Indeed, the lack of clearly defined rules of community living and ambiguous decision making were among the primary causes of the collapse of the intentional communities of the 1960s–1970s (Christian 2015). Most communitarian experiences of this 'hippie wave' disappeared because they were unable to manage the gap between highly idealistic aspirations and the organisation of daily life.

Thus, when systems of shared governance began to emerge and became refined over the past 70 years, communities were able to take up the challenge of combining concrete parameters (land occupation, construction, regulations, housing, food, activities providing individual and/or collective income) with similar values (social justice, open education, equality, participation). As such, the wave of eco-villages in the late 1990s has been their first major beneficiary; the observed longevity of the eco-villages studied (those operating for five to 30 years) seems to confirm that by giving governance bodies a significant role, more mature and stable communities have taken root over time.

Another historical perspective reminds us that the desire for a more inclusive, egalitarian, and participatory society has been one of the strongest motivations behind communitarian initiatives since the architectural and social utopias of the 19th century. The ecological vision has always been supported by the same values of equality, democracy, and solidarity (Audier 2017) that we must use in order to halt the negative impacts of human activity on the environment and repair the damage we have caused. In other words, we must return to the true sources of wealth: *frugal* life economies that are rooted in family and social life. The clearly stated desire for reconnection with nature, and more broadly with the living world (from systemic to telluric) remains the timeless key to the articulation between the various communitarian waves, from the first communities of the 19th century to today's eco-villages.

Beyond visible spatial forms and interactive social flows, a third level of space production exists, even if it is more intangible. Eco-villages are not limited to resolving paradoxes between ecological values and socio-economic constraints in particular. Eco-villages also emerge as privileged laboratories for resolving tensions, whether technical, social, or existential. Some of the researchers cited above (Beltrán Gutiérrez 2024; Brown et al. 2019; Berardi et al. 2020) argue that empathy opens up 'an ability to understand another's plight and metaphorically to "walk in their shoes" ' (Brown et al. 2019), thus becoming the foundation of sincere and supportive social interactions.

Empathic interactions heighten awareness of tensions and support the development of a form of collective life that is continuously adaptive. The pursuit of a balance between individual and collective well-being is embedded in the shared governance systems observed in our study, which recognise the often contradictory coexistence of individual and collective needs. Potential tensions that might hinder the harmonious development of the community are identified and sometimes addressed preventively through communication practices (such as overcoming prejudice), cooperative processes (dialogue, listening), and participatory modes of decision making (committed presence). In this sense, empathy functions as a vector of human connection that supports the development of the communitarian project in both its social and spatial dimensions.

Through these structured and inclusive frameworks, eco-villages strengthen their capacity to prevent and resolve internal conflicts while increasing their resilience to external tensions: 'Empathy is the tool we must use as individuals and communities to not only rebuild, but thrive and advance successfully for years to come' (Brown et al. 2019).

If empathy fosters a climate of exchange that allows for the production of soft and fluid forms of Inhabiting, we still need to pay attention to the many obstacles to conflict resolution. Such conflicts can even affect the sustainability of the eco-village, including psychological and emotional barriers, divergent opinions on the evolution and adaptation of the communitarian project, economic issues, and the societal and cultural factors of each individual living in the community.

Forms of empathic Inhabiting in networks, from the local to the global

The examples from the communities cited have made it possible for us to grasp that the language of communitarian architecture begins with the spatialisation of subsistence activities, and that it takes shape in specific places (whether outdoors or indoors, collective or shared). This organisation is articulated through various layers of functioning (subsistence economies and non-closed territorial anchoring, shared governance and empathy-driven dynamics). In practice, empathy would be a 'relational toolbox' enabling eco-villages to produce *soft* (or *low-tension*) forms of Inhabiting by establishing points of 'respectful porosity' between the spheres of economy and ecology, between producing domestic-social space and Inhabiting a territory, and between daily interactive flows and democratic values. This approach to forms of communitarian Inhabiting leads us to examine how the spatial forms produced are situated in the history of architecture and how local eco-villages participate in the formation of a planetary network.

From local integration to global standardisation

Let us briefly recall that since the 1920s, the field of architectural history has considered that the aesthetic canons of modernity in architecture are structured around principles such as the purity of forms, functional rationality, and the innovative use of industrial materials such as concrete and steel. In other words, *form follows function*, an architectural principle attributed to the American architect Louis Sullivan (1856–1924). Inspired by the Modern Movement, notably the architects Le Corbusier (1887–1965) and Ludwig Mies van der Rohe (1886–1969), these principles gradually integrated postmodern critiques in the 1960s and 1970s, particularly those of Robert Venturi (1925–2018) and Aldo Rossi (1931–1997).

Alternative currents began to emerge simultaneously in the 1960s as a reaction to the aesthetic canons of modernism and anti-modernism. These approaches favoured organic, irregular, soft, and flexible forms as well as natural and recycled materials. They were also linked to critical regionalism (Frampton 1983) and vernacular architecture (Rudofsky 1964; 1977) that emphasised local traditions and human-scale architecture. Two 20th-century architects, Hassan Fathy (Paquot 2009) and Paolo Soleri (Soleri 1969), embodied this trend by emphasising ecological sustainability and harmonious integration into the landscape as opposed to the *functional uniformity of modern architecture* (Figure 8.3).

The period spanning from the 1960s to the present day clearly highlights an evolution in architectural language between the back-to-the-land movement (1965–1985) and the emergence of eco-villages in the mid-1990s.

FIGURE 8.3 Hallingelille eco-village, Denmark

Many members of the hippie communities of the 1960s–1980s were mostly artist-architects or engineer-tinkerers. They initiated alternative construction methods and forms, what can be described as 'signal' or 'performative architectures' that asserted functional and stylistic approaches that were diametrically opposed to the aesthetics canons of architectural modernism. They explored the construction of geodesic domes, pioneered the integration of recycled materials (car bodies, aluminium or glass cans, and used tires), and rediscovered the art of building huts.

Just as they significantly matured in their modes of governance, eco-villages born in the 1990s have developed architectural productions that now draw on a much more traditional vocabulary. This aesthetic shift translates into the construction of individual houses or very small collective buildings made of local and ecological materials (wood, straw, and hemp) that complied with local urban planning regulations (which was not always the case of the builders of so-called hippie communities). However, geodesic domes, reinterpreted and significantly improved in terms of energy performance, can still be found in eco-villages here and there.

For instance, at the Danish eco-village of Hallingelille, several families have chosen the geodesic dome and zome (a hybrid structure combining domes and other space-filling shapes) forms that implicitly reference this countercultural shape while applying ecological construction standards to it and subdividing the open space. These choices also recall certain traditional dwellings of the Faroe Islands, an autonomous archipelago that is part (or could be considered as a region) of the Kingdom of Denmark.

Existing buildings in eco-villages are often restored by builders inspired by traditional local construction know-how. This is the case of the farmhouse converted into a common house at Le Hameau des Buis or the incredible rehabilitation project undertaken by the founders of Torri Superiore. It took this eco-village no less than 20 years of participatory construction workshops and the residents' energy to rebuild a medieval *castello* with only a few necessary adaptations. Following the village house or its variant, it reflects the scale of the hamlet, which has been commonly adopted in the communitarian world. Like the vast majority of current eco-village projects, La Cité Écologique, Le Hameau des Buis, and Hallingelille were designed from the outset with interlocking domestic perimeters: a common house and shared gardens at the centre, a few dwellings nearby, then surrounding farmland or forest operations.

The shift from an 'architecture of protest' to an 'architecture integrated into territories' undoubtedly reflects the eco-villages' intention to blend into their host environment. Compared to communities from the back-to-the-land movement, their architecture deliberately embraces a less experimental aesthetic but one that would be more acceptable to neighbours. Consequently, the village house signals the gradual disappearance of the geodesic dome as the emblematic form of countercultural communities. But in return, it introduces much better ecological performance (natural materials, energy consumption, level of neutrality, and/or autonomy of constructions).

While the desire to integrate into territories can lead to a trend toward uniformity, the architectures reflect a sense of belonging to the larger 'global eco-village' family. The similarity of forms and spatial scales observed in La Cité Écologique located in Quebec, Canada and New Hampshire in the US, Hallingelille in the Copenhagen region of Denmark, and Le Hameau des Buis in Ardèche in France, is particularly visible. In line with the previously described 'eco-logical' organisations, the aesthetic choices remain linked to the desire of a sustainable social project driven by a harmonious (empathic) social dynamic connected to a local territory. Thus, the form of communitarian Inhabiting generates another specific 'space of porosity': at the heart of their building approaches, the sites observed connect activities of voluntary simplicity and degrowth with alternative currents of architecture.

Let us return here to Van der Ryn's vision, which is rare, if not unique, in the field of architectural design. From his point of view, the designers' capacity for empathy goes hand-in-hand with that of future users. Empathy is seen as the key to a transformative behavioural dynamic capable of proposing sensitive constructions that interact with all living systems. These practices have become spontaneous in the communitarian world, where it is not surprising to observe that self-construction is very common in the creation sites of eco-villages (collective workshops, participatory work camps). The builders of these communities often avoid the advice of 'conventional' architects, with whom they consult only for administrative formalities.

The movements of self-construction and vernacular architecture are historical examples (Ravéreau/ALADAR 2013; Paquot 2009) that are rarely addressed even today in the education and practice of architecture. Although marginal, these examples raise some of the most cutting-edge contemporary issues in eco-architecture, including participatory construction, material recycling, economy of means and space, integration of renewable energy, and techniques adapted to geography and climate.

Because they are open and demonstrative experiences, eco-villages could constitute a true corpus of time-tested forms and techniques for future architects, i.e., bioclimatic construction, use of local materials, promotion of traditional know-how, and the care that must be taken to harmoniously integrate dwellings into their natural and social environments. The empathic dimension required for genuine integration remains a 'sensitive bridge'. This bridge is strategic: it can enable connections between human activities and ecological contexts in ways that respect their integrity. However, when a situated awareness of place is not considered in its deeper implications, the establishment of genuine integration may fail, affecting the development of the eco-village—particularly in its relationships with neighbouring territories and in its initial interactions with the outside world. Here, we can better understand the subtle bonds that must be woven and maintained for ecological spatial organisations to function sustainably (Figure 8.4).

Eco-villages and the world map

After having explored the local scale, let us now examine the global geographies of communitarian Inhabiting, where a shift—both intentional and unprecedented in the history of communitarian movements—takes place from local settlement to a globally organised network. The history unfolding from the 1960s to the present shows that the communitarian quest in all its diversity is not limited to the idea of living in ecological dwellings to pursue a rural life disconnected from the rest of the world. Communitarian life has always been connected to its place of settlement. Its history speaks of population movements between regions and continents (Petitfils 2011; Creagh 2009), and it has also always been concerned with reciprocal exchange and the building of interaction networks.

At this scale, the fields of architecture and technological development (Picon 2013) appear deeply interwoven. This observation has been reinforced

FIGURE 8.4 Le Hameau des Buis eco-village, France

over the past 20 years by a new type of connection: the Internet both shapes and displays the presence of the communitarian world on the global map. Like the majority of documented initiatives, the eco-villages cited here are affiliated with national or international organisations that have gained remarkable momentum thanks to digital technologies. The two most unifying platforms today are GEN (Global Ecovillage Network) and FIC (Foundation for Intentional Communities). This level of international networking is unprecedented; it is a significant marker of the transformations initiated and experimented with by the current eco-villages.

To Patrick Geddes' trilogy, *Place Work Folk*, which preceded and inspired the bioregionalism and territorialism movements, we must now add the word *Net*. This suggests that we consider the framework *Place Work Folk Net* as an update to the layers of interaction and data that need to be addressed. While rooted in geographic locations with unique physical characteristics (Places) hosting real-life experiences (Work, Folk), the virtual (Net) resonates with the global map of eco-villages. Local-global connections are now limitless; geographical boundaries and concepts of distance become blurred. The Internet demonstrates that living in an eco-village means choosing a micro-local human scale of settlement while fully benefitting from a multi-scalar globalised world.

This unprecedented technological openness may well continue to foster a renewed expansion of the global communitarian movement. In France, early studies analysing the effects of the COVID-19 pandemic on urban exodus show that while major urban centres remain central, the migration from cities to rural areas has become steady, with signs of acceleration suggesting that older patterns of movement are deepening (PUCA 2022). The rapid evolution of this phenomenon, which is influenced by today's affordable urban housing problem, will require more time in order for us to draw long-term conclusions about the current patterns of migration.

However, the momentum represented by current eco-villages points to an unparalleled rise in the diffusion of experiments, inter-communitarian meetings, and the production of content, particularly in response to the need for exchange during lockdown periods and the broader impacts of the health crisis, including video conferences, gatherings, symposia, and training. A more ecological world at every scale is, of course, at the centre of these discussions. Without exception, communities and eco-villages are connected to and active in these 'new geographies of globalization' (Lussault 2017) through websites, social media, webinars, and directories or maps located on international platforms. As multi-dimensional experiences, contemporary communities and/or eco-villages demonstrate a fundamental tension in their history: they must strike a balance between connection and disconnection even though their rural lifestyle choices are deeply aligned with a project grounded in Nature and empathetic relationships with all forms of life. It is important to note that this tension distinguishes eco-villages from virtual communities and that their projects are materially embodied in architectures and spatial arrangements that can be observed.

Conclusion

Let us imagine for a moment a geography of the world devoid of empathy. Would it not result in impoverished territories and spaces that disregard human interaction, ecosystems, and socio-cultural or ethical contexts? Then consider the forms of communitarian Inhabiting. Even though the study summarised here is necessarily partial, it appears that eco-village experiments address complex equations on a daily basis that involve Inhabiting across different territorial scales (from local to global) without endangering the planetary habitat.

Is there not, in fact, great potential for tension in the simultaneous acts of breaking away from modern urbanity, embracing voluntary simplicity, wanting more autonomous lives, inventing democratic governance models, and building a viable daily economic system? Without the 'relational toolbox' that empathy represents, wouldn't these inherent tensions in any alternative project tip into conflict and ultimately lead back to mainstream patterns that ignore the finitude of natural resources or the disintegration of the community? Instead, we observe that eco-villages are increasingly numerous, resilient, and long-lasting, and that they contribute to the preservation and regeneration of the Earth's biotope for all species and for future generations (Noriega 2024; De Burca 2025).

While eco-villages spontaneously implement empathic design, they also offer insight into how both internal tensions of daily life (such as non-violent communication and participatory governance) and external pressures (such as territorial cooperation and environmental regeneration) can be constructively addressed. Eco-villages are not idyllic spaces but living laboratories where the tensions inherent in systemic interactions are resolved through ongoing negotiation and adaptation processes.

Two further arguments open up the architectural interpretation of communitarian spaces to network-based practices. On the one hand, the formal evolution of eco-villages' architectural productions demonstrates a link to alternative architectural currents with a strong ecological dimension, including vernacular forms and know-how, self-construction, and collective building practices. On the other hand, as locally rooted yet globally connected entities, eco-villages are capable of carving out specific paths through the vast global exchanges that emphasise the transmission of knowledge and values shared by the communitarian world.

These empathic forms of knowledge and values undoubtedly contribute to ecological transversality, i.e., the ability to link everyday practices, spatial organisation, and ecological concerns across different levels. They might also have the potential to be shared more broadly and perhaps even to inspire the fields of architecture and contemporary society as a whole.

Finally, one of the most significant assets of eco-village initiatives is that they are situated at the opposite end of the spectrum from 'global solutions', which are unlikely to succeed in overcoming ecological challenges. The more realistic hopes for genuine solutions to our ecological problems now lie in reconstituting solidarity-based micro-regions that are capable of harmoniously reinterpreting urban-rural relations (Paquot 2022).

In this sense, eco-villages represent a valuable historical and contemporary resource that should inspire architectural education and practice for at least two reasons: firstly, the diversity of communities as tangible long-term experiments rooted in site-specific adaptation and secondly, the invention of specific ways of life that integrate ecological regeneration processes with local ecosystems, thereby shaping the planetary conditions inherited by future generations.

The capacity of eco-villages to intuitively integrate the empathic design principles described by Sim van der Ryn into lived and site-specific architectures makes them particularly relevant in times of ecological urgency. These community-based experiences operate on multiple fronts simultaneously. On the one hand, they inspire current reflections on architectural education by contributing to a transformation in ways of conceiving and building with intergenerational effects. On the other hand, they open up concrete pathways for informing local policies, particularly with regard to land use and the consideration of existing ecosystems.

More broadly, and beyond these situated contexts, the principles of empathic design implemented on a daily basis by intentional communities and eco-villages can inspire both contemporary architecture and territorial planning, provided that the necessary shift toward ecological and solidarity-based forms and scales of planning is embraced.

References

ALADAR. (2013). André Ravéreau: Biographie. André Ravéreau | ALADAR. https://andreravereau.org/andre-ravereau/biographie/

Alexander, C., Ishikawa, S., & Silverstein, M. (1977). *A pattern language: Towns, buildings, construction*. Oxford University Press.

Amar, S. (2020). *Laboratoires d'architectures écotopiques: Des communautés d'hier aux écovillages d'aujourd'hui (États-Unis–Europe, 1965–2015) [Ecotopian architecture laboratories: From yesterday's communities to today's ecovillages (United States–Europe, 1965–2015)]*. ENSA-Marseille, Laboratoire INAMA, Aix-Marseille Université [ENSA-M (National School of Architecture of Marseille), INAMA Laboratory; AMU (Aix-Marseille University)].

Audier, S. (2017). *La société écologique et ses ennemis: Pour une histoire alternative de l'émancipation [The ecological society and its enemies: Toward an alternative history of emancipation]*. La Découverte.

Beltrán Gutiérrez, M. T. (2024, 10 October). *Empathy design*. Anhalt University of Applied Sciences. www.researchgate.net/publication/384803711

Berardi, M. K., White, A. M., Winters, D., Thorn, K., Brennan, M., & Dolan, P. (2020). Rebuilding communities with empathy. *Local Development & Society*, 1(1), 57–67. https://doi.org/10.1080/26883597.2020.1794761

Brown, K., Adger, W. N., Devine-Wright, P., Anderies, J. M., Barr, S., Bousquet, F., Butler, C., Evans, L., Marshall, N., & Quinn, T. (2019). Empathy, place and identity interactions for sustainability. *Global Environmental Change*, 56, 11–17. https://doi.org/10.1016/j.gloenvcha.2019.03.003

Christian, D. L. (2015). *Creating a life together: Practical tools to grow ecovillages and intentional communities*. New Society Publishers.

Creagh, R. (2009). *Utopies américaines: Expériences libertaires du XIXe siècle à nos jours [American utopias: Libertarian experiments from the nineteenth century to the present]*. Agone.

De Burca, J. (2025). *Les meilleurs éco-villages du monde [The best ecovillages in the world]*. Constructive Voices. https://constructive-voices.com/fr/les-meilleurs-%C3%A9co-villages-du-monde/

Frampton, K. (1983). Towards a critical regionalism: Six points for an architecture of resistance. In H. Foster (Ed.), *The anti-aesthetic: Essays on postmodern culture* (pp. 16–30). Bay Press.

Joubert, K. A., Dregger, L., & Global Ecovillage Network (Eds.). (2015). *Ecovillage: 1001 ways to heal the planet*. Triarchy Press.

Leroux, N. (2008). Qu'est-ce qu'habiter? [What does it mean to inhabit?]. *VST – Vie sociale et traitements [VST – Social Life and Care]*, 97, 14–25. https://doi.org/10.3917/vst.097.0014

Lockyer, J., & Veteto, J. R. (Eds.). (2013). *Environmental anthropology engaging ecotopia: Bioregionalism, permaculture, and ecovillages*. Berghahn Books.

Lussault, M. (2017). *Hyper-lieux: Les nouvelles géographies politiques de la mondialisation [Hyper-places: The new political geographies of globalization]*. Éditions du Seuil.

Magnaghi, A. (2017). *La conscience du lieu* [*The consciousness of place*]. (P. Larochelle, S. Larochelle, & E. Bonneau, Trans.). Eterotopia France.

Magnaghi, A., & Bonneau, E. (2014). *La biorégion urbaine: Petit traité sur le territoire bien commun* [*The urban bioregion: A short treatise on territory as a common good*]. Eterotopia France.

McGinnis, M. V. (1998). *Bioregionalism*. Taylor & Francis.

Noriega, A. (2024, 13 June). *Communautés durables: L'essor des écovillages à l'échelle mondiale [Sustainable communities: The global rise of ecovillages]*. Driving ECO. www.drivingeco.com/fr/les-communaut%C3%A9s-durables-explosent-les-%C3%A9covillas-au-niveau-mondial/

Paquot, T. (2005). Habitat, habitation, habiter. *Informations Sociales* [*Social Information*], 123, 48–54. www.cairn.info/revue-informations-sociales-2005-3-page-48.htm

Paquot, T. (2009). Hassan Fathy, construire avec ou pour le peuple? [Hassan Fathy: Building with or for the people?]. *Cahiers d'histoire. Revue d'histoire critique* [*History Notebooks: Journal of Critical History*], 109, 15–25. https://doi.org/10.4000/chrhc.1907

Paquot, T. (2022). L'espérance biorégionale [Bioregional hope]. *Topophile*. https://topophile.net/a/10689

Paquot, T., Lussault, M., & Younès, C. (2007). *Habiter, le propre de l'humain: Villes, territoires et philosophie.* [*Inhabiting, the essence of the human: Cities, territories, and philosophy]*. La Découverte.

Petitfils, J.-C. (2011). *Les communautés utopistes au XIXe siècle* [*Utopian communities in the 19th century*]. Pluriel.

Picon, A. (2013). Learning from utopia. *Journal of Architectural Education*, 67(1), 17–23. https://doi.org/10.1080/10464883.2013.767120

PUCA – Plan Urbanisme Construction Architecture [Urban Planning, Construction, Architecture]. (2022, 17 February). *L'exode urbain? Petits flux, grands effets: Les mobilités résidentielles à l'ère (post-Covid)* [*Urban exodus? Small flows, major effects: Residential mobilities in the (post-Covid) era]*. Report. Ministère de la Transition Écologique [French Ministry for Ecological Transition]. www.urbanisme-puca.gouv.fr/l-exode-urbain-petits-flux-grands-effets-les-a2388.html

Rudofsky, B. (1964). *Architecture without architects: A short introduction to non-pedigreed architecture*. Museum of Modern Art.

Rudofsky, B. (1977). *The prodigious builders: Notes toward a natural history of architecture*. Secker & Warburg. https://archive.org/details/prodigiousbuilde00rudo

Soleri, P. (1969). *Arcology: The city in the image of man*. Organism Earth Library. www.organism.earth/library/document/76

Van der Ryn, S. (2013). *Design for an empathic world: Reconnecting people, nature, and self*. Island Press.

Van der Ryn, S., & Cowan, S. (2010). *Ecological design* (10th ed.). Island Press.

Wright, E. O., Farnea, V., & Peschanski, J. A. (2017). *Utopies réelles* [*Real utopias*]. La Découverte. (Original work published 2010.)

Notes

1 In this article, I use 'Inhabiting' with a capital 'I' to signal its expanded conceptual meaning as both a spatial practice and a value-based mode of existence.

2 The eco-villages that are not part of this study are: The Lama Foundation (since 1967, New Mexico, US), Arcosanti (since 1970, Arizona, US), Earthship (since 1994, New Mexico, US), and Ham-Nord-CAN (since 2003, New Hampshire, US).

Bibliography

Angers Magazine. (2023, 28 August). Pourquoi les écovillages sont la solution pour changer de vie en faisant avec moins et avec l'autre [Why ecovillages are a solution for changing one's life by living with less and with others]. *Angers Magazine*. www.angersmag.info/pourquoi-les-ecovillages-sont-la-solution-pour-changer-de-vie-en-faisant-avec-moins-et-avec-lautre_a11846.html

ARE – Office fédéral du développement territorial [Federal Office for Spatial Development]. (1987). *Le rapport Brundtland* [*The Brundtland Report*]. ARE. www.are.admin.ch/are/fr/home/medien-und-publikationen/publikationen/nachhaltige-entwicklung/brundtland-report.html

Carson, R. L. (1962). *Silent spring*. Fawcett.

Ehrlich, P. R. (1971). *The population bomb*. Ballantine Books.

Global Footprint Network. (n.d.). *Ecological footprint*. Global Footprint Network. www.footprintnetwork.org/our-work/ecological-footprint/

Heidegger, M. (1973). *Essais et conférences* [*Essays and lectures*]. (A. Préau, Trans.; J. Beaufret, Pref.). Gallimard.

Jonas, H. (1992). *Le principe responsabilité: Une éthique pour la civilisation technologique* [*The imperative of responsibility: In search of an ethics for the technological age*]. (J. Greisch, Trans.; 2nd ed.). Éditions du Cerf. (Original work published 1979).

Journée des écovillages [Ecovillages Open Day]. (n.d.). La Cité Écologique. https://citeecologique.org/fr_CA/nous-visiter/journee-des-ecovillages/

Kraus, S. (2012). *Patrick Geddes, architecte de paysages et médecin de l'environnement* [*Patrick Geddes, landscape architect and physician of the environment*]. Metagraphies. http://metagraphies.org/Sir-Patrick-Geddes/EXPO-PG_09-12.pdf

Lamoureux, J. (2023). Immersion d'un jour: Est-ce que les écovillages peuvent changer le monde? [A one-day immersion: Can ecovillages change the world?] *Urbania*. https://urbania.ca/article/immersion-dun-jour-est-ce-que-les-ecovillages-peuvent-changer-le-monde

Latouche, S. (2011). *Petit traité de la décroissance sereine* [*A short treatise on serene degrowth*]. Mille et une Nuits.

Meadows, D. H., Meadows, D. L., & Randers, J. (1972). *The limits to growth: A report for the Club of Rome's project on the predicament of mankind*. Universe Books.

Naess, A. (2013). *Écologie, communauté et style de vie* [*Ecology, community, and lifestyle*]. (C. Ruelle & H.-S. Afeissa, Trans.). Éditions Dehors.

Paquot, T. (2015). *Désastres urbains: Les villes meurent aussi* [*Urban disasters: Cities die too*]. La Découverte.

Ravéreau, A., Bertaud du Chazaud, V., & Ravéreau, M. (2007). *Du local à l'universel* [From local to universal]. Editions du Linteau.

Truong, N. (2020, 18 August). Écovillages, logements participatifs...Une autre manière d'habiter la Terre [Ecovillages, participatory housing... Another way of inhabiting the Earth]. *Le Monde*. www.lemonde.fr/series-d-ete/article/2020/08/18/ecovillages-logements-participatifs-une-autre-maniere-d-habiter-la-terre_6049180_3451060.html

United Nations Educational, Scientific and Cultural Organization (UNESCO). (n.d.). *Fonds d'urgence pour le patrimoine, 2023: Rapport annuel d'activité [Emergency Fund for Heritage, 2023: Annual Activity Report]*. UNESCO Bibliothèque numérique [UNESCO Digital Library]. https://unesdoc.unesco.org/ark:/48223/pf0000391415_fre

9

SUSTAINABILITY BROKERING AND GREEN COMMUNITY BUILDING IN DANISH SOCIAL HOUSING

Freja Bach Kristensen

Introduction

'Climate crunch time is here' (United National Environment Programme, UNEP 2025: XI). The United Nations' (UN) 2025 Emission Gap report makes it abundantly clear. In order to stay within the 1.5-degree pathway for global warming, we need worldwide mobilisation on a scale and at a pace never seen before. Current strategies and nationally determined contributions are not enough. At the same time, we see an increasing public dissatisfaction with climate policies across Europe (Tallent 2025). Governments are balancing climate mitigation efforts with considerations towards perceived social equity and justice issues. In Denmark, Lars Aagaard, the Danish Minister for Climate, went so far as to say in 2023 that climate policy initiatives should be supported by 80% of the population (Nielsen 2023). In response to this challenge, public policies, organisation strategies, and grassroots initiatives addressing the transition to a less fossil fuel-dependent future increasingly point to *communities* as key to enabling green transition initiatives (Taylor Aiken, Eadson, Hobson, & Dinnie 2022; Agyemen, Schlosberg, Carven, & Matthews 2016; Fischer, Holstead, Hendrickson, Virkkula, & Prampolini 2017).

In the latest report from the Intergovernmental Panel on Climate Change (IPCC), civil society social movements are advanced as a primary driver of social and institutional change (Shukla et al 2023: 556). The report highlights the ability of, for example, the non-governmental organisation (NGO) Transition Towns and community energy projects to 'reconcile lower carbon footprints with higher life satisfaction and higher incomes', which helps

DOI: 10.4324/9781003769064-10

mitigation initiatives gain social acceptance (Shukla et al 2023: 551–556). In the Danish context, we see community building as a climate governance instrument. Danish municipalities, in particular, are promoting community building as a tool to foster climate friendly behaviour among its citizens to help meet carbon reduction targets.

However, the expectation for communities that are praised for being small-scale and local to create system-wide impact introduces instrumental reasoning onto and into community work (Taylor Aiken, Eadson, Hobson & Dinnie 2022: 4,6). This risks shifting attention away from local community aspirations and needs. These purpose-driven communities thus face a dilemma about how to simultaneously attend to community needs and local desires while contributing to alleviating societal issues. The promotion of local communities in solving the climate crisis might also be read as a neoliberal governance strategy of pushing governmental responsibilities onto citizens (Hobson 2002: 102; see also Winter 2019).

With the increased strategic interest and investment in green community formation, we need to explore and understand how purpose-driven community formation actually takes shape. By analysing a participatory case study of green community building in social housing in Denmark, this chapter explores how an outside push for community formation is attended to and navigated by residents. The study is based on 11 months of fieldwork in a social housing estate on the outskirts of Copenhagen.

Residents living in social housing generally have a lower carbon footprint than the rest of the population because of their lower incomes and fewer square metres of housing.[1] However, there are several reasons why it is still relevant to study green community building in this sector. Firstly, the social housing sector in Denmark houses almost 20% of the population. This means that even small reductions can have a big impact nationally. Secondly, because the sector houses a substantial part of the lowest socio-economic groups, it is a relevant site to understand conceptualisations of just and equitable transitions.

My research found that promoting social green activities per se is not always sufficient to build liveable green communities in social housing estates. The success of green communities is dependent on the engagement of brokers, key individuals who take on the practical and mental labour of green community building. I use the concept of *sustainability brokering* to unfold the role of these individuals and how they engage in translations of the wider notion of sustainability into locally attuned activities that can mobilise other residents.

This chapter presents the cases of two women, Anne and Gitte, who are residents in the social housing estate. I show how local sustainability brokers are essential for successful green community building in social housing and discuss how this work can be supported and promoted. I argue, firstly, that it is imperative to make room for aligning green community building with individual goals and life projects. Secondly, that sustainability brokering requires not only an understanding of the sustainability agenda, social and organisational skills, and a legitimate position within the local community, but also the ability and room to manoeuvre in order to address other pressing local issues and agendas. Thirdly, the chapter shows that sustainability brokering is never an isolated endeavour. It only works when there is collaboration and support from those with access to strategic agendas and resources.

Before presenting the main findings, the chapter begins with a walkthrough of the role of communities in the green transition, followed by the theoretical framework, an introduction to the case study, and an outline of the methodology.

The role of communities in the green transition

Community is increasingly seen as essential to the advancement of the green transition, both within social movements and in public policies (Taylor Aiken, Eadson, Hobson, & Dinnie 2022; Agyemen, Schlosberg, Carven, & Matthews 2016; Fischer, Holstead, Hendrickson, Virkkula, & Prampolini 2017). This community orientation presents a shift in previous approaches to climate governance. Dating back to the first UN report on sustainable development from 1987, commonly known as the Brundtland Report, sustainability is often defined as economic and social development that meets the needs of the present without compromising future generations' ability to meet theirs (WCED 1987: 17).

As described by cultural geographer Kersty Hobson, this definition of sustainability has been adopted primarily as a *rationalisation* paradigm in state policies, where the focus is on doing more with less (Hobson 2002; Hobson 2013). Within this framework, sustainable development is primarily linked to efficiency and technological innovation. Citizens are incentivised to change consumer practices based on information campaigns and economic regulation that call on the rational and efficient citizen-consumer to stimulate market supply (Hobson 2013: 1086).

As Elizabeth Shove (2010) argues in her influential critique of the so-called 'ABC approach' (Attitudes, Behaviours, Choices), strategies for climate change mitigation that target individual human behaviour are, however, not enough to create significant societal change. Instead, Shove argues, we should look towards the social and collective organisation of practices as the relevant analytical and social setting for understanding and influencing behaviour change (Shove 2010). The orientation towards communities and social relations as central for social and institutional change reflects this understanding. But is it also a shift away from the rationalisation paradigm in climate governance?

In a Danish context, the orientation towards community in climate governance can be found in Danish municipalities in particular. Their work with climate mitigation is anchored in official climate action plans. In the forthcoming revisions of these plans, the scope is broadened to include targets for reducing consumption-related carbon emissions (Concito 2025: 30). This change in scope means that municipalities may need to influence citizens' consumption patterns in order to meet their climate targets. This is politically sensitive because it touches upon individual lifestyle choices and ethics.

To address this political balancing act, several Danish municipalities have established a partnership with the Green Neighborhood Communities NGO. For example, the municipality of Brøndby, a suburb of Copenhagen, has hired a local organiser to act as a liaison between the municipality and local groups of volunteers who work with climate initiatives. This organiser also helps more people engage in climate-friendly activities on the premise that 'everyone needs to contribute to the green transition, and we need to do it together' (author's translation, Brøndby Municipality 2022: 4). Brøndby's overall strategy is to encourage citizens to make climate-friendly choices by supporting local green communities, using libraries and culture houses as inspiration hubs and providing the necessary infrastructure for sharing, repairing, and recycling (Brøndby Municipality 2022: 7).

This strategy in Brøndby municipality reflects the understanding that behaviour is embedded in shared social practices, as proposed by Shove (2010) and others (Hargreaves 2011; Shove & Spurling 2013; Strengers & Maller 2014). On the face of it, this approach seems to be fundamentally different from the rationalisation paradigm focused on targeting the individual citizen-consumer through information campaigns and economic incentives. However, the moral call for citizens to care about the environment runs the risk of framing sustainable consumption as a personal responsibility and duty for the good environmental citizen, which reflects neoliberal governance strategies of responsibilisation (Hobson 2002: 102; see also Prothero, Claudy, & Keating 2025; Winter 2019). As a stand-alone strategy, this approach follows the same pathway as the rationalisation paradigm: it avoids addressing the underlying issues of overconsumption and continuous economic growth and pushes the responsibility onto individuals instead of implementing unpopular political regulation and systemic changes.

However, community building as a way to advance green transition initiatives doesn't necessarily follow the rationalisation paradigm. As outlined in the introduction to this book, green communities offer an important place to re-imagine ways of living by offering alternative visions for the future. Green communities are highlighted for their documented ability to challenge the presumed connection between growth and life satisfaction, which is contrary to the win-win understanding of reconciling lower carbon footprints with higher life satisfaction and higher incomes put forward by the IPCC.

Community is praised as a prominent location for realising structural change and future liveability in both academia and beyond, which reflects a *radical* sustainability paradigm focused on systemic change.

As noted by Taylor Aiken et al. (2022), regardless of which sustainability paradigm is invoked, this instrumentalisation of community means 'that the community shifts from a relational and present purposiveness towards a more rational and demonstrable way to conceive of and justify a community's tasks'. (Taylor Aiken, Eadson, Hobson & Dinnie 2022: 4). The community's value becomes displaced; its effects are located and legitimised elsewhere. The authors are critical of this instrumentalisation of grassroots community movements, which, they argue, devalues the non-scalable elements of community such as face-to-face encounters and place-specific activities (Taylor Aiken, Eadson, Hobson & Dinnie 2022: 6).

However, purpose-driven community building as a concept and a practice is growing; it is being driven by both civil society movements and public policy. Rather than dismissing these communities as neo-liberal governance tools or as detached from their local setting, we need to understand how people navigate and negotiate different understandings of sustainability and seek to build meaningful relations around it. As I demonstrate in the following sections of this chapter, in the context of green community building in a Danish social housing estate, key individuals assume this position of translating and mobilising across different contexts. First, I introduce the theoretical framework of this analysis.

Theoretical framework – Community building and sustainability brokering

Community is a notoriously difficult concept. It is invoked to signify anything from close-knit relations to national identity but is rarely explicitly defined (Amit 2010; Amit & Rapport 2002). To shed light on the formation of purpose-driven communities, I employ Finnish anthropologist Pilvi Posio's (2024) conceptualisation of community as *collective aspiring*. This concept was developed through her studies of long-term recovery after the 2011 tsunami and Fukushima nuclear plant disaster in Japan. The term focuses specifically on future orientation as the formative element of community. It brings our attention to the ways in which community is actualised and enacted through social practices and events where individuals consciously or unconsciously strive to align desired futures coming together around a shared goal (Posio 2024: 378, 383).

It is important to note here is that this is not a process in which everyone can participate on equal terms as it takes place within existing structural conditions and hierarchies, as Posio highlights (Posio 2024: 387). Posio further reminds us that 'the social dynamics of collective aspiring involves the power to impose desired futures on others by structuring social practices and peoples' actions and temporalities within them' (Posio 2024: 387).). When looking at the formation of purpose-driven communities, it is thus important to understand how desired futures are aligned and how the capacity to aspire is shaped and negotiated. Here I find the analytical figure of the broker useful.

While the broker has been portrayed historically as a mediator between two encapsulated political structures (the village and the nation, for example) (Lindquist 2015), current literature on brokering focuses on the more generative capacity of this role. One example shows how tourist guides who are active in producing and negotiating specific readings of cultural heritage become co-constructors of not only social groups but also of abstract concepts such as authenticity (Macdonald 2006). Others have shown how brokers are crucial to the emergence of new collectives by establishing new links between formerly unconnected sites or groups such as local land right disputes and larger global paradigms of indigenous rights (Bräuchler 2021).

When we look at the strategical and instrumental desire to establish green communities, I find it relevant to explore how these political intentions take shape. How can we understand them as places of environmental socialisation, that is, places where 'new pro-environmental social identities, interactions, and relations are forged'? (Hargreaves 2011: 96). In the context of green community building as a climate governance tool, I am interested in how these top-down sustainability strategies and policies *become real* (Mosse & Lewis 2006: 13). David Mosse and David Lewis argue that in order to do this, we need to draw attention to the social life of policy and the making of development worlds by institutionally and socially embedded individuals (Mosse & Lewis 2006: 13). They define these individuals as brokers simultaneously engaged in the 'maintenance of coherent representations of social realities and in the shaping their own social identities' (Mosse & Lewis 2006: 16). Instead of approaching brokering as a mediation between predefined entities and hierarchical scales, they frame brokering as an ordering process carried out by individuals. Thus, brokering becomes an act of assembling, that is, bringing together actors, institutions, and resources to make a temporary unity (Koster & van Leynseele 2018: 804).

This and other work sheds light on how certain individuals play a key role in brokering the often abstract aspirations of policies with the lived realities of the target populations to give life to the conceptual intentions in strategies and documents (see, for example, Chalhi, Koster, & Vermeulen 2018). Likewise, with the concept of sustainability increasingly being written into municipal and organisational strategies for social housing, I trace how these political ambitions are given life through 'translation' by key local residents. I use the concept of sustainability brokering to describe how these local brokers, by combining their own personal life projects, circumstances, and skillsets with the opportunities arising from strategically appointed resources, translate the idea of environmentally sustainable living into a lived reality and mobilise other locals to join green activities, and hereby assemble more or less temporary green communities.

I am not the first to coin the concept of sustainability brokering. To shed light on how more sustainable variants of energy practices can be promoted, Russel Hitching introduces sustainability brokering in his work on energy use (Hitchings 2013). He uses the concept to refer to both the act of redefining conventions towards more sustainable standards within an existing community of practice and bringing a new community of shared practices into being by developing and promoting situations where people might share ideas and talk about less resource-intensive ways of living (Hitchings 2013: 11). His conceptualisation of sustainability brokering underlines the co-constitutive process of community building and the implementation of green or sustainable actions by directing our attention to how practices are formed and negotiated in a forum of peers.

However, the term also presents sustainability as a predefined realm of appropriate conduct that different actors can tap into to renegotiate ways of doing certain everyday actions. It does not address how a sustainable practice is conceptualised in the first place and what would encourage anyone to take on the role of introducing it. By approaching sustainability brokering as an act of assemblage, I show how sustainability as a concept is brought to life by local sustainability brokers. In the following section, I introduce the background and details of the case study.

A case of green community building in the Danish social housing sector

Following the trend in the field of climate mitigation, 'Ambitious Green Communities' is one of the strategic goals for 2024–2026 of BL, the national association for Danish non-profit housing organisations. The focus is to provide information on how to make climate- and environmentally friendly decisions in the management of estates and among their residents (BL 2023: 8). This goal is echoed in the strategic plans of local housing organisations and estates, not least in AAB, the oldest and largest housing association in Denmark and the site of my fieldwork. In 2018, the association formulated its overall vision as 'providing the frames for living the good life in a beautiful, healthy and environmentally friendly residence at a reasonable rent' (author translation, AAB 2018: 1), with a strong emphasis on supporting local communities in the individual estates.

Since 2024, AAB has committed to using the platform 'AlmenKompas', which is an Environmental, Social & Governance (ESG) tool developed specifically for the social housing sector in Denmark. Under the three themes, it offers specific indicators that can assist in monitoring the sustainability efforts in the individual housing organisations and estates. AAB describes this commitment as founded in the recognition of their impact and social responsibility as a social housing association with a yearly turnaround in the billion kroner class (AAB 2025: 4). AAB further emphasises their commitment to sustainability in their construction policy. Here, considerations about the environmental impact of construction are accompanied by the intention that construction should support community efforts and involve the residents as much as possible in the development of their own area and the estate (AAB 2021: 3).

In the case of the specific housing estate, Estate 49, where I conducted fieldwork, this commitment was supported by an application to the government Agency for Culture and Palaces for a six-year art project that could supplement the physical renewal plan that had been initiated. In the following section, I introduce the estate and present some details about the social housing sector in Denmark in order to give us some context.

Introduction to Estate 49

The first thing that meets the eye when arriving at Estate 49 is the four 8-storey concrete high-rise buildings connected by bright blue covered walkways. These buildings stand out significantly from the four 4-storey buildings on the estate and the surrounding housing stock of mainly 3–4-storey brick and mortar apartment blocks. The estate houses approximately 600 residents and was designed by Danish architect and declared functionalist Gunnar Milthers. It was built with prefabricated concrete elements in 1967–1969. Milthers was a strong advocate for high-rise construction: he argued that it could substantially cut costs and provide more open green spaces for residents (Milthers 1954). The estate is typical for its time and showcases what others have labelled ‘welfare landscapes’ that materialise ‘the Danish welfare state in concrete, asphalt, plantings, and earthworks’ (Jessen & Tjetjen 2021: 475). Unlike its equivalent in other countries, the social housing sector in Denmark is first and foremost a *public* housing sector. The intention has always been to provide affordable housing for a broad segment of the population, and the first Danish housing policies advancing social housing constituted a cornerstone in the establishment of the country’s welfare state (Bech-Danielsen 2022: 401).

The story of Estate 49 follows the general history of the sector. After the building boom in the 1960s and 70s when one-third of all Denmark’s social housing was built, the sector experienced a collapse. Due to the interplay of economic developments and construction problems, the larger modernist housing estates built for the working-class family experienced a concentration of disadvantaged residents. Despite significant efforts over the years to solve both physical and social problems on such estates, since the 1980s these housing areas have housed the largest proportion of people without a connection to the labour market, the largest proportion of people with the lowest income and education levels, and the highest share of people with multiple criminal convictions (Weatherall, Boje-Kovacs, & Esgaard-Pedersen 2016: 10). These areas have also experienced a significant increase of residents with an immigrant background, leading to the popularisation of the word ‘ghetto’ in Danish public debates about social housing during the 2010s.

Issues of ethnicity and integration have increasingly taken centre stage since the early 1990s, culminating with the passing of The Parallel Society Act in 2018. This law, commonly known as ‘the ghetto law’, requires disadvantaged housing areas categorised as ‘parallel societies’ to develop extensive physical renewal plans that reduce the number of social housing family units to 40% of accommodations offered, either through demolition or densification. The categorisation is based on the residents’ connection to the labour market, education level, income level, conviction rates, and the proportion of residents who are immigrants from non-Western countries or their descendants (The Government of Denmark 2018).

Grouped together with two nearby housing estates, Estate 49 was categorised as both a ‘vulnerable housing area’ and a ‘prevention area’ over the years because of the residents’ higher conviction rates and low average income compared to national and regional averages, respectfully. Today, the area no longer figures on any such list because the resident population has decreased to under 1000 residents. This is partly due to the consolidation of units, which is part of the estate’s physical renewal plan that was approved by the residents in March 2019. The necessity of the renewal plan is framed as both meeting renovation needs, such as bringing the sewage system up to date, and as a crime prevention measure (Copenhagen Municipality 2020). The estate has a high tenant turnover rate, which is partly connected to the larger number of small housing units (36–58 m^2). The plan calls for 98 of these units to be consolidated into 48 larger family units.

Another main feature of the plan is a so-called infrastructure connection. Currently, the estate only opens out to the surrounding neighbourhood on one side of the triangular area, a common feature of modernist-inspired housing estates. The intention with the infrastructure connection is to enable a passage through the estate that opens it up to non-residents. The hope is that this, among other measures such as improved lighting, can help address feelings of being unsafe that were reported by approximately 24% of the residents in an internal survey from 2018.

Construction to implement the details of this plan is expected to last over three years and also entails the building of a new community house. The art project named 49 Open Doors that was granted by the Agency for Culture and Palaces began prior to the construction work with the primary goal of co-creating six permanent art pieces to be part of the estate. The goal is that the art project will act as a catalyst for community development and strengthen social cohesion in the estate, which would honour the AAB's trifold commitment to sustainability. Besides the housing association and the two artists involved, the local library and several social work initiatives are also part of the project, which currently operates from a small shed that functions as a temporary community house. While the project itself does not have a specific sustainability agenda that addresses environmental issues, it is generally occupied with supporting local engagement and place identity.

I contacted the project in the fall of 2023, and with the approval of the project manager, Heidi, and the local board of tenants, I was allowed to work from the temporary community house and collaborate with the other partners in the art project. My plan to introduce green activities was welcomed by both the board of tenants and the project partners. I was specifically tasked with revitalising the local community garden, but I also ended up establishing a handicraft group, doing smaller events, and attending other activities in the estate.

Methodological approach

The case study was carried out as a participatory study as part of the larger collaborative research project SAMSKAB. The goal of this project is to support the development of new green communities and understand how participation in green communities affects environmental behaviour and contributes to social cohesion. Pilot money was set aside to help initiate and support green community activities.

Because I faced navigating the tension of coming into the field with a declared intention, I took inspiration from Caroline Gatt and Tim Ingolds's concept of *correspondence*, an approach that shifts the emphasis from participant observation to *observational participation* (Gatt & Ingold 2013: 154). This entails the anthropologist's deliberate and reflexive participation in the ongoing happenings in the field (Gatt & Ingold 2013: 141). As described in length elsewhere (Bach Kristensen, forthcoming), I found this approach allowed me to bring the intentions and resources of the project as well as my own personal skillset in dialogue with the aspirations and concerns of the residents and other actors working in the sector.

Fieldwork was conducted from November 2023 to October 2024 and also took place in another estate. My empirical material consists of written and recorded field notes as well as 18 taped interviews with 16 residents and 11 professionals working in the sector. Research material also include pictures and documents such as meeting minutes, organisation diagrams, plans for event management, and more produced by me and other actors in the field. The material was categorised into relevant tasks and approaches to find similarities and differences across actors. The initial analytical categories were further explored by talking with the residents during interviews and in a subsequent workshop that eventually led to the development of the overarching conceptualisation of the sustainability broker.

Sourcing a sustainability broker

When I first started my fieldwork in Estate 49, I wrote a brief introduction of myself that was posted on the internal Facebook site and hung on a wall in the community house. I asked if anyone was interested in engaging in green activities and listed examples such as a repair café or communal eating. I also announced a weekly 'open house' in the monthly calendar that was distributed to the residents. But no one showed up. It seemed like no one else wanted to engage in the type of sustainable activities I envisioned.

One day, however, I received a phone call from Heidi, the coordinator of the local community building project. She told me that there was a resident who was very interested in green activities at the estate, and she wanted to say hi. I quickly jumped on my bike and left the university in the central city to greet the short-haired, middle-aged woman whom I came to know as Gitte. As we sat on the benches outside the shed on a sunny afternoon, we quickly got into a discussion of the possibility of attaching a rainwater collection system to the temporary community house. It was surrounded by planters with herbs and flowers, and Gitte thought it would be a better way to provide water for the plants as well as for birds and other wildlife. Heidi hoped it would remind the residents to water the plants and encourage feelings of ownership. I was mostly just excited to get an activity started.

Realising this project, however, proved to be more difficult than I had first imagined. Neither the project coordinator nor Gitte owned a car, which made it both difficult and expensive to collect second-hand materials, which all three of us thought was the sustainable approach. I said that I could provide a car if Gitte could find the materials, but after a few texts back and forth, our efforts stalled. Eventually, Heidi ordered new supplies and the resident artists in the project helped Gitte and Heidi install them. A water collection container and a few watering cans were then placed outside the shed.

Now, I rarely saw anyone other than Gitte water the plants, but installing the water collection system proved the beginning of a fruitful collaboration. As the months passed, Gitte came to play a major role not only in the green community activities I was involved in but in the community building project in general. This wasn't a given from the outset. Initially, Gitte had been very frustrated about the project. The shed was placed right outside her apartment window and blocked her view of the green space and its vibrant birdlife. After several conversations with her, this frustration turned into engagement. In her own words, Gitte got involved so she could influence the project so everyone could tolerate it, which to her included a larger focus on sustainability and biodiversity.

For Gitte, sustainability was not articulated as being about either efficiency or systemic change, but rather as an overall dissatisfaction with the general state of the world. Her interest in sustainability came from a lived experience of having to be creative with few resources and a broader informed concept of climate change. She considered making use of existing resources and taking care of what you already have as simple common sense. This unfolded through practical suggestions and initiatives, for example, in the community garden I was trying to revitalise. Here Gitte would come up with different projects such as using leftover cobblestones to create a garden bed or building bird baths out of different second-hand materials. We frequently talked about ideas, and I assisted with sponsoring and sourcing materials.

However, not all the other residents saw Gitte's common-sense perspective of taking care of things around you as, well, commonsensical. One of the projects Gitte was working on was making soapbox cars out of discarded prams and other recycled materials. She felt it was important to find something for the kids to do. The playgrounds in the estate have seen better days, and Gitte understood why the kids might start to think about less constructive activities when they had nothing better to do. I frequently saw kids racing down the little hill in the estate shrieking with joy as the soapbox cars caught momentum. However, she would also confront the kids when they didn't act in what she considered an appropriate way. One day, Gitte scolded some of the kids for climbing on the temporary community house and hitting it with a stick. This led to a conflict with the parents and the ethnic group they were part of, since they did not take it well that Gitte had corrected their children's behaviour.

Gitte and I also had our disagreements. One day, she called me on the phone upset about what was happening in the community garden. 'If it is going to be like this', she told me, 'I don't want to be part of it'. She was referring to the fact that we had cut down some of the bushes during a workday in the garden. Gitte is much occupied with the wildlife nearby, especially the porcupines that somehow manage to inhabit the small green areas in the urban housing estate. She was very adamant that the gardens should foster biodiversity and create habitats for wildlife. We managed to talk through it. I explained to her that it was important to some of the other residents that the garden looked presentable but that no one was interested in making a streamlined garden scape.

Gitte also had a conflict with one of the other project partners at a joint workshop in the project where we were building birdhouses for the estate. The project partner was very focused on how the activity could engage residents and less on the actual practical outcome. This meant, to Gitte's frustration, that the size of the holes in the birdhouses did not correspond to the type of bird it was supposed to attract. Sustainability aspirations and community building efforts do not always align seamlessly.

Regardless, Gitte's engagement was crucial to the enabling of community activities. As the months passed, the project partners became more mindful of how to spend their resources. Creating and facilitating activities for the residents required more than expected. For example, someone needed to turn up the heating in advance of reading club meetings in the winter months. Besides coming up with ideas and offering her practical expertise, these were the kind of things Gitte would assist with. But she did question whether her contributions were properly recognised and compensated. She often commented to me that her friends questioned whether she was being 'used' by the project. In the end, Gitte's continued commitment and contributions to the community building project led to a small paid position to help manage the area and support community activities. (This position has since been extended by the local Board of Tenants even though Gitte is due to move out of the estate).

Gitte played a crucial role in both channelling resources into concrete relevant sustainability activities and supporting community efforts, although the two didn't always align. There were, however, other cases where the integration of sustainability considerations and community building was less contested, as the story of Anne shows.

Brokering as an act of translation

On a warm and sunny summer day, we are gathered in the little wooden trailer that serves as the temporary community house. Anne, an energetic, quick-to-laugh, middle-aged woman, is showing me and six others, mostly elderly women in the handicraft club, how to use plastic bags as a material for crocheting. It isn't easy to get the chatty crowd to be quiet, but Anne has a natural authority about her and manages to get everyone's attention. She introduces the different techniques that can be used to cut the plastic into strips and explains what the width of the strip means for the finished product and how it can be combined with yarn to make a more durable result.

She also shows us different items she has crocheted: a coaster, a cover for an empty pill container that now serves as a little wallet, and a beautiful wide-brimmed summer hat that we all try on in turn. We joke about whether it would be suitable for a rainy day or if the holes would still make your hair wet. All the women are very curious and eager to try out the new techniques while Anne moves around the room giving us advice and gently sparring with us when we encounter difficulties. She manages to make room for the usual chitchat, suggestive jokes, and laughter that characterises the handicraft club meetings. A week later, one of the women shows us a basket she has made from plastic bags from a local supermarket. It now sits on permanent display in the little trailer.

Because of my initial frustrations about not being able to gather anyone around green activities, I had begun to abandon the concept of an explicit green agenda. After talking to some older residents about a handicraft club that was active in the estate in the 1970s and 1980s. I took the initiative to revive it. Handicraft can be about repair and resource awareness, I thought. So, I issued an invitation to a start-up meeting and after that, the handful of women who attended and I began to meet every other week. I was explicit about my goal: I told the women that my research project was about green communities and tried to introduce different elements that I conceived of as sustainable such as plant milk and repairing clothes.

But it wasn't until Anne joined us that the sustainability angle really took off. I had been a little hesitant about pushing the green agenda. Anne, on the other hand, had plenty of initiative, such as organising the workshop on how to repurpose leftover plastic bags as a material for crocheting.

Reflecting back on the workshop, I later asked Anne what it requires to take on an authoritative role in a community of peers. 'It has to be fun', Anne told me, 'and feel natural. Or else it will create an awkward atmosphere'. Having finished my fieldwork at the time of this interview, I asked Anne if the topic of upcycling and the like is still something they talked about in the handicraft club. 'Yes, it is', she told me. Just earlier that day, they had been sharing small Christmas gifts and Anne had wrapped her presents in recycled wrapping paper. After seeing her beautifully wrapped presents, the other women started collecting the leftover wrapping to try to reuse it. 'So it rubs off', she says. Anne's entry in the handicraft club was instrumental in transforming it to a forum for discussing and promoting a less resource-intensive lifestyle.

Anne's own interest in sustainability had come on gradually. It was borne out of necessity during a period of her life where she had a limited income. But it also came from a profound curiosity about how to repurpose things and an overall concern with the current state of the planet and what she termed the hopeless practice of overconsumption. However, her desire to introduce new recycling-oriented practices in the handicraft club wasn't only about changing consumption habits. She was planning to open an online shop to sell her creations and offer different crochet patterns and workshops. So, she also saw the handicraft club as a forum to test out her ideas and refine the format for her workshops. Furthermore, the handicraft community gave her a better understanding of what was going on in the estate and was also a direct link to the chair of the board of tenants, who was a regular attendee.

When I had to end my fieldwork, Anne agreed to take over the coordination of the handicraft club. It wasn't a comprehensive task, but it required coordinating with the project manager in charge of the temporary community house about the monthly calendar. Since several of the older women were not very tech-savvy, Anne would need to be available on SMS if someone wasn't attending. She also needed to welcome new participants, participate in larger events at the estate, such as the summer party, and apply for funding by the board of tenants after I could no longer provide pilot project funding. Although these duties might seem minor, doing them were vital to the functioning of the handicraft club and to the wider community. Anne played a key role in sustaining the community and introducing the sustainability agenda.

Supporting a sustainable brokering practice

A central challenge emerges when we look at community building as a governance tool for furthering the green transition. How do we balance community needs and what is desirable in the local context with achieving the broader effects of reducing carbon emissions or enabling system-wide change? The case study of Estate 49 shows how the sustainability agenda unfolds on the ground in emerging green communities and yields important lessons.

From introducing new more sustainable materials in the handicraft club to showing creative ways to recycle materials and creating awareness about biodiversity and resource use, the cases of Anne and Gitte show that it is possible to gather people around less resource-intensive activities and inspire more environmentally friendly behaviour. It does, however, as argued by Hitching (2013: 108), require a sensitivity to the community and its conventions. As Anne explained, engaging a community of peers around green activities requires certain skills, including making it fun. This not only requires being sensitive to the community's specific social codes and its sense of humour but also the ability to communicate well and take on a role of authority.

However, merely inviting people to engage in green activities does not initiate or guarantee community building. Mobilising neighbours around green activities requires continuous practical and mental labour, such as sourcing materials, coming up with suitable ideas, and creating occasions and spaces where people can meet and form relationships. Building local green communities is not just the result of strategic decisions and funds. To engage and mobilise locals around a sustainability agenda, someone needs to make it suited for and attuned to the local context.

However, this might take forms that are not anticipated or do not fit into organisational expectations. As the case of Gitte demonstrates, local sustainability concerns might clash with project intentions and it might even be necessary to sideline the green agenda to make room for local interests, as I did with the handicraft club. Neither Gitte's nor Anne's approach to sustainability represented specific paradigms; they were largely built on their experiences of making a living with few resources but good creative skills. Sustainability brokering always entails this personal dimension.

Brokers are often described as serving their own interests by trying to maintain their positions as those controlling the flow of resources (Baron 2021: 65). However, as political scientist Steffen Jensen puts it, the position as broker is not always strategic or even wilful: 'More often, the intention may simply be to make a life that is meaningful right here and right now, not reducible to a function of a system or a resistance to it' (Jensen 2018: 891). Anne takes on the role of sustainability broker because there is a complementarity between the community building work and her own personal ambitions. Through her skilful social navigation and ability to align the resources available with her own interests, she managed to assemble a green community. Gitte's involvement was based on trying to make the project tolerable to the residents, and her engagement eventually led to a paid position. Securing this position not only required a huge investment of voluntary time on Gitte's part, it involved recognition by the project manager and the housing estate that the time and presence offered by Gitte was absolutely necessary for the continued success of the community's activities

As noted by Brauchler et al. (2021), brokering is not only an individual endeavour, it takes part through chains of brokerage, i.e., connections between individuals from the strategic to the grassroots levels. Gitte's employment was only possible because the project manager cut back on her hours. I also supported Gitte's sustainability ambitions when she encountered difficulties in navigating the estate's social relations and she helped me realise the strategic ambitions of the research project.

I have focused in this chapter on the individual residents, but several people play a role in channelling funds and translating specific conceptual policies into local realities. In the case of Estate 49, Heidi, other project partners with their wider organisational backdrop, and I came in with both agendas and funds. For purpose-driven community building to be successful in social housing estates, it is necessary to understand the interlinked nature of sustainability brokering and how those closer to decision making and the allocation of funds can collaborate with the residents to create successful sustainability initiatives.

Conclusion

In this chapter I have introduced the concept of sustainability brokering to explain the role played by individuals in engaging their neighbours and building local communities around green activities. Based on a specific lived experience of resource scarcity, creative skills, and an intimate understanding of the lived social life in the estate, specific conceptualisations of sustainability were brought to life in communal activities. These conceptualisations focused on general resource awareness and creative and quality-of-life-enhancing social practices such as making art of recycled materials, gardening, building soapbox cars, and crocheting with used plastic bags.

Being able to engage in these acts of brokering required the skilful navigation and practical commitment of the individual sustainability brokers to channel project resources into community activities that both resonated with their own goals and aspirations and were attractive to the other residents who joined in. Even though the agenda of green community building is being supported increasingly by municipalities, national organisations, and housing associations, there is not always recognition of the locally adapted efforts that bring these strategic ambitions to life.

I argue that succeeding in purpose-driven community building requires an understanding and accommodation of the individual motivations and goals of the brokers. It also requires a flexible approach towards how local sustainable actions might unfold and an inclusive approach towards other local agendas. Finally, I argue that sustainability brokering is never performed in isolation. It unfolds through a chain of individuals, from those developing sustainability strategies and setting carbon reduction goals through project managers and to the local residents who bring these strategic ambitions to life. My research has focused on the last link in the chain, the residents. Further research is needed to shed light on brokers at each level.

However, one thing stands out clearly. For purpose-driven community building to be successful in social housing estates, those who are responsible for sustainability strategies and green project development should think of ways to recognise and compensate the work being done locally by people in less advantageous situations. This can be done by delegating parts of the budget to local salaries or finding other ways to give back and compensate the work of sustainability brokers in translating the wider notion of sustainability into collective local green activities that can help foster a less resource-intensive lifestyle.

Note

1 The average income for residents in social housing is 33% lower than the average of the rest of the population (The Danish Building Fund 2025: 53). On average, social housing residents occupy 21% fewer square metres, which equates to 42.9 m^2 per resident (The Danish Building Fund 2020: 1).

References

AAB. (2018). Boligforeningen AAB – en forening for stærke lokale fællesskaber. Mission, vision og strategi. [Housing Association AAB – an association for strong local communities. Mision, vision and strategy]. *AAB*. https://aab.dk/wp-content/uploads/2022/06/AABs-vision-mission-og-strategi.pdf

AAB. (2021). Bæredygtige fællesskaber – AAB's byggepolitik. [Sustainable communities – AAB's construction policy]. *AAB*. https://aab.dk/wp-content/uploads/2021/07/Byggepolitik.pdf

AAB. (2025). Årsrapport 2024. Bæredygtighed og Sociale indsatser i AAB. [Annual report 2024. Sustainability and social initiatives in AAB]. *AAB*. https://aab.dk/wp-content/uploads/2025/06/AAB_AaRSRAPPORT-2024.pdf

Agyeman, J., Schlosberg, D., Craven, L., & Matthews, C. (2016). Trends and directions in environmental justice: From inequity to everyday life, community, and just sustainabilities. *Annual Review of Environment and Resources*, 41, 321–340. https://doi.org/10.1146/ annurev-environ-110615-090052

Amit, V. (2010). Community as 'good to think with': The productiveness of strategic ambiguities. *Anthropologica* 52(2), 357–363. www.jstor.org/stable/29546038

Amit, V., & Rapport, N. (2002). *The trouble with community: Anthropological reflections on movement, identity and collectivity*. Pluto Press. https://doi.org/10.2307/j.ctt18mvnx3

Bach Kristensen, F. (forthcoming). Making it happen: Green community building in Danish social housing. In S. Krøijer, I. Hougaard, & S. Breslin (Eds.), *The process of impact: Doing and documenting change anthropologically*. Palgrave.

Baron, R. (2021). Cultural brokerage revisited. *Journal of Folklore Research*, 58(2), 63–104. https://doi.org/10.2979/jfolkrese.58.2.03

Bech-Danielsen, C. (2022). Ideology of the welfare state in solidified concrete for housing. In K. M. Raahuage, D. Simpson, M. Søberg, & K. Lotz (Eds.), *Architectures of dismantling and restructuring: Spaces of Danish welfare, 1970-present*. Lars Müller Publishers.

BL. (2023). Almene Mål 2024–2026. [Social housing goals 2024–2026]. *BL*. https://bl.dk/media/cqxhfg0q/bl_almene_maal_20242026.pdf

Bräuchler, B., Knodel, K., & Röschenthaler, U. (2021). Brokerage from within: A conceptual framework. *Cultural Dynamics*, 33(4), 281–297. https://doi.org/10.1177/09213740211011202

Brøndby Municipality. (2022). Klimaplan for Brøndby – vejen mod et CO2-neutralt og klimatilpasset fællesskab. [Brøndby climate plan – the road towards a CO2-neutral and climate adapted community]. https://brondby.viewer.dkplan.niras.dk/plan/81#/

Chalhi, S., Koster, M., & Vermeulen, J. (2018). Assembling the irreconcilable: Youth workers, development policies and 'high risk' boys in the Netherlands. *Ethnos*, 83(5), 850–867. https://doi.org/10.1080/00141844.2017.1362452

Concito. (2025). *Status på kommunernes klimahandling – Etablering af et monitoreringssystem*. [Status of the municipalities' climate action – Establishing a monitoring system]. Klimaalliancen.

Copenhagen Municipality. (2020). Bilag 2 Renovering af almene boliger 1. halvår 2020. Appendix 2. Renovation of social housing 1. Half of 2020]. www.kk.dk/sites/default/files/agenda/59366ee6-dbbf-4594-9a03-25bc5f3232fb/eabefe81-67c0-4c38-9caf-b4fac2fe4a46-bilag-2.pdf

Fischer, A., Holstead, K., Hendrickson, C. Y., Virkkula, O., & Prampolini, A. (2017). Community-led initiatives' everyday politics for sustainability – Conflicting rationalities and aspirations for change? *Environment and Planning A: Economy and Space*, 49(9), 1986–2006. https://doi.org/10.1177/0308518X17713994

Gatt, C., & Ingold, T. (2013). From description to correspondence: Anthropology in real time. In W. Gunn, T. Otto, & R. C. Smith (Eds.), *Design anthropology: Theory and practice* (pp. 139–158). Bloomsbury.
Hargreaves, T. (2011). Practice-ing behaviour change: Applying social practice theory to pro-environmental behaviour change. *Journal of Consumer Culture*, 11(1), 79–99.
Hitchings, R. (2013). Sharing conventions: Communities of practice and thermal comfort. In E. Shove, & N. Spurling (Eds.), *Sustainable practices: Social theory and climate change*. Routledge
Hobson, K. (2002). Competing discourses of sustainable consumption: Does the 'rationalisation of lifestyles' make sense? *Environmental Politics*, 11(2), 95–120. https://doi.ord/ 10.1080/714000601
Hobson, K. (2013). 'Weak' or 'strong' sustainable consumption? Efficiency, degrowth, and the 10-year framework of programmes. *Environment and Planning C: Government and Policy*, 31(6), 1082–1098. https://doi-org.kb-ku.idm.oclc.org/10.1068/c12279
Jensen, S. (2018). Epilogue: Brokers – pawns, disrupters, assemblers? *Ethnos*, 83(5), 888–891. https://doi.org/10.1080/00141844.2017.1362455
Jessen, A., & Tietjen, A. (2021). Assembling welfare landscapes of social housing: Lessons from Denmark. *Landscape Research*, 46(4), 474–494. https://doi.org/10.1080/01426397.2020.1808954
Koster, M., & van Leynseele, Y. (2018). Brokers as assemblers: Studying development through the lens of brokerage. *Ethnos*, 83(5), 803–813. https://doi.org/10.1080/00141844.2017.1362451
Lindquist, J. (2015). Brokers and brokerage, Anthropology of. In *International encyclopedia of the social & behavioral sciences* (2nd ed.). Elsevier. https://doi.org/10.1016/B978-0-08-097086-8.12178-6
Macdonald, S. (2006). Mediating heritage: Tour guides at the former Nazi Party Rally Grounds, Nuremberg. *Tourist Studies*, 6(2), 119–138. https://doi-org.ep.fjernadgang.kb.dk/10.1177/1468797606071473
Milthers, G. (1954, 15 October). Utraditionelle Højhuse i Rødovre. [Untraditional highrises in Rødovre]. *Arkitekten* [*The Architect*]. Weekly Edition. https://danskbyggeskik.dk/Publikationer/1714%20-%20Utraditionelle%20h%C3%B8jhuse%20i%20R%C3%B8dovre.pdf
Mosse, D., & Lewis, D. (Eds.). (2006). *Development brokers and translators: The ethnography of aid and agencies*. Kumarian Press Inc.
Nielsen, S. B. (2023, 11 August). Klimaminister: 'Vi skal have en klimapolitik, som 80 procent af danskerne kan se sig i' [Minister for Climate: 'We need a climate politics that 80 percent of Danes can recognize']. *Danmarks Radio (Danish Broadcasting Cooperation)*. www.dr.dk/nyheder/politik/klimaminister-vi-skal-have-en-klimapolitik-som-80-procent-af-danskerne-kan-se-sig-i
Posio, P. (2024). Aspired communities: Reconsidering community in light of the temporal dimensionality of social life. *Anthropological Theory*, 24(4), 371–396. https://doi.org/10.1177/14634996241231667
Prothero, A, Claudy, M., & Keating, A. (2025). Beyond neoliberalism: The role of community in the responsibilization of citizen-consumers during the great recession. *Journal of Consumer Research*, 2025, ucaf039. https://doi.org/10.1093/jcr/ucaf039

Shove, E. (2010). Beyond the ABC: Climate change policy and theories of social change. *Environment and Planning A: Economy and Space*, 42(6), 1273–1285. https://doi.org/10.1068/a42282

Shove, E., & Spurling, N. (Eds.). (2013). *Sustainable practices: Social theory and climate change* (1st ed.). Routledge. https://doi.org/10.4324/9780203071052

Shukla, P.R., Skea, J., Slade, R., Khourdajie, A., van Diemen, R., McCollum, D., ... Malley, J. (Eds.). (2023). *Climate change 2022: Mitigation of climate change. Contribution of working group III to the sixth assessment report of the intergovernmental panel on climate change.* Cambridge University Press. doi: 10.1017/9781009157926.

Strengers, Y., & Maller, C. (Eds.). (2014). *Social practices, intervention and sustainability: Beyond behaviour change* (1st ed.). Routledge. https://doi.org/10.4324/9781315816494

Tallent, T. (2025). A green divide? Climate policy support and its rural geography in Europe. *West European Politics*, 2025-06, 1–31. https://doi.org/10.1080/01402382.2025.2521589

Taylor Aiken, G., Eadson, W., Hobson, K., & Dinnie, L. (2022). Vicarious scale instrumental imaginaries in community sustainable transitions. *Global Environmental Change*, 75, Article 102543. https://doi.org/10.1016/j.gloenvcha.2022.102543

The Danish Building Fund. (2020). *Temastatistik 2020:4. Boligforbruget i den almene boligsektor 2010-2019* [*Thematic statistics 2020:4. Housing consumption in the social housing sector 2010–2019*]. The Danish Building Fund.

The Danish Building Fund. (2025). *Statistik. Beboere i den almene boligsektor 2024* [Statistics. Residents in the social housing sector 2024]. The Danish Building Fund.

The Government of Denmark. (2018). *Aftale mellem regeringen (Venstre, Liberal Alliance og Det Konservative Folkeparti) og Socialdemokratiet, Dansk Folkeparti og Socialistisk Folkeparti om: Initiativer på boligområdet, der modvirker parallelsamfund* 9. maj 2018 [*Deal between the government (Venstre, Liberal Alliance, The Conservative People's Party, The Social Democrat Party, The Danish Peoples' Party, and The Socialist Peoples' Party) about: Initiatives in the housing sector to prevent parallel societies*]. The Government of Denmark.

United Nations Environment Programme (UNEP). (2024). *Emissions Gap Report 2024: No more hot air ... please! With a massive gap between rhetoric and reality, countries draft new climate commitments.* UNDP. https://doi.org/10.59117/20.500. 11822/46404

Weatherall, C. D., Bøje-Kovács, B. J., & Egsgaard-Pedersen, A. (2016). *Et historisk tilbageblik på de særligt udsatte boligområder udpeget i 2014: Udviklingen i tilflyttere, fraflyttere og fastboende* [*A historical review of the disadvantaged housing estates designated in 2014: Trends in in-migration, out-migration and long-term residents*]. Kraks Fond Byforskning.

Winter, A. K. (2019). The green city citizen: Exploring the ambiguities of sustainable lifestyles in Copenhagen. *Environmental Policy and Governance*, 29(1), 14–22. https://doi.org/10.1002/eet.1837

World Commission on Environment & Development (WCED). (1987). *Our common future: Report of the World Commission on Environment and Development.* Oxford University Press.

10

SLOW PRACTIVISM

Volunteering in Dutch urban gardens

Cristina Grasseni

Introduction

Urban gardens exist in most European cities, often on municipal land situated at the margin of cities. Gardening volunteers thus depend on the goodwill of the municipal governments to access land, or on non-governmental organisations (NGOs) and their projects, when mandated by local governments. This chapter discusses the ambivalences that sometimes characterise the interaction between volunteers—interested in learning new skills and experiencing practices of 'grassroots responsible innovation' (Grasseni 2018)—and management agendas, which often impose, apparently by economic necessity, top-down budget cuts and austerity narratives that stifle volunteers' motivations.

This chapter first describes in detail a case study of an unnamed city in the Netherlands. Then, I broaden the discussion to other European cases, developed in a comparative project that I recently led. The overall argument is that the social, environmental, and political values and skills found in communities of food, craft, and practice is much broader than the economic value gained from food production per se. Failing to recognise the 'alternative' (namely, other than financial) value of these communities of practice, grafts irreconcilable expectations onto urban food gardens. Focusing on the skills and the relevance of craft practice to future-making, I compare and contrast the everyday practice of cultivating depleted urban soil in densely populated regions in the context of urban agriculture in the Netherlands. I argue that the expectation that urban agriculture is a sustainable innovation that should quickly bring about system change can be hegemonic; it can impose

DOI: 10.4324/9781003769064-11

timescales and criteria for efficiency and productivity that do not belong in either volunteering or food growing.

Based on fieldwork conducted in the Netherlands with food horticulturalists since 2020, I consider how self-managing, learning, and social participation go hand-in-hand with discourses of sustainability. Defining an appropriate scale and pace of activities includes rediscovering and celebrating taste, seasonality, and pleasure in food preparation and conviviality, where the materiality of food, tools, spaces, and rhythms plays an important part. However, volunteer communities impact sustainability politics to a lesser degree. Market and entrepreneurial economics remain the orthodox frame of reference for policy making and for the management of volunteering projects and organisations.

In scholarship, volunteering in food gardens has sometimes been read as implicitly falling short of more radical positions relating to food justice and just transition (Agyeman et al. 2003). It can also be associated with self-centred concerns of consumer-activists about reclaiming local knowledge and control over food production (Stanton 2017). To complement this analysis, I offer a socio-cultural perspective that acknowledges the collective aspect of self-production. I stress that collective social learning in food gardens is a form of slow resistance that is experiential rather than ideational.

It is no surprise that social gardens in large cities are very popular with visitors, since they offer a wealth of social and health benefits. Some of these 'green' services tangibly increase social and relational, and well-being (Hankins 2023). For example, some gardens are organised by non-governmental organisations (NGOs) in collaboration with social and mental health services, so that access to such gardens under supervision is subject to medical prescription (for example, as a result of burnout). These 'healthcare farms' (*zorg boerderijen*) are reimbursed by the national health service for daycare-allocation of patients (*dagbesteding*). In addition, pandemic and climate crises have revealed the vulnerability and inequality of socio-economic systems based on global commodity flows that make consumers dependent on few and far between centres of material production and knowledge creation. In the face of logistical collapse, the issue of global de-skilling and local re-skilling has become more evident. Alternative provisioning, including food procurement, can be a part of re-skilling. Collective action in green communities, however, may be impeded by issues of scale (when the scale of a project can be deemed too small or too exclusive) or by the lack of skills to re-engineer the supply chain (Grasseni 2024).

I present here ethnographic insights that draw on my volunteering at a 'city farm' (*stadsboerderij*) in a Dutch city. This is a historical farm comprised of animal sheds, a heritage orchard and pasture meadows, with later additions of a café and food shop, garden patches for educational gardening, and a building for administrative offices, workshops, assemblies, and celebrations.

This space functions as a socialisation hub where paid employees, volunteers, customers, and visitors can meet and interact with each other.

The chapter further draws on the results from *Food Citizens?*, a research project funded by the European Research Council (ERC). Its title includes a question mark to indicate scepticism about the generalised application of the notion of 'food citizenship'. The project compared collective food procurement networks in three representative cities in Europe—Gdańsk, Rotterdam, and Turin—to contrast and juxtapose local categorisations of solidarity, diversity, skill, and scale (see www.foodcitizens.eu). In my definition, collective food procurement comprises networks of people who connect with each other around practices of self-production, exchange, direct sale, and local food policy. Our research addressed which considerations of solidarity and diversity grew, if at all, among and within these networks, the scale of their practices, and how they approached and valued skills (for example, gardening skills or organisational skills in volunteer work, or in crafting food). The project also highlighted the paradoxical contradictions of 'food citizenship' as a normative discourse for grassroots practice, highlighting the broad, sometimes conflicting range of engagements it encompasses.

In the following ethnographic analysis, I ask how and if, given competing interests and power dynamics, food volunteers are caught between hegemonic understandings and expectations of active citizenship and their own grassroots forms of practical engagement with working, learning, and producing food. I use this as a departure point to argue for what I call *slow* practice.

The skills of quiet practivism

Through the ethnography that I conducted over the years, I have explored multiple forms of skilled practice—first in dairy farming, then in cheese-making in the Italian Alps, in alternative food networks in Italy, solidarity economy networks in Massachusetts (USA), and last, in food gardening in the Netherlands (Grasseni 2025). Across these disparate economic practices, people appropriate notions and practices of learning and of activism. In the following analysis, I use theories on the education of attention, apprenticeship, and craft to analyse the political, moral, and ecological positions of these 'green' communities (see Gausset & Jensen 2024).

Discussions on skills, i.e., enskillment (Ingold 2000, 2013), re-skilling (Stead & Coulton 2022; Sekules 2020), and de-skilling (Herzfeld 2007) in the contemporary post-industrial 'ruins' of late capitalism (High et al. 2017) leave us at a crucial turning point. On the one hand, in post-industrial, late-capitalist countries, generational change is an issue for the transmission of the skills of procuring and processing food. On the other hand, there is an abundance of creative approaches to food procurement.

For example, according to Joe Smith and Petr Jehlička (2013), 'quietly sustainable' gardeners in post-socialist Central and Eastern Europe maintain sustainable lifestyles within localised food systems, continuing the tradition of keeping a garden patch and exchanging, swapping, distributing, and gifting harvested food. Unlike in Western Europe and North America, where community gardening is seen as an alternative act of resistance, Poland's urban food gardens, for example, benefit from well-organised gardening associations that for decades have allocated urban land to factory workers' families in the expectation that they contribute to meeting essential food needs (Gracjasz 2024). Also, as research in the Czech Republic shows, food gardeners reap additional health and environmental benefits from the food's freshness and quality, which is celebrated in terms of the taste of food, as well as the benefits of safety nets that are nourished by the exchange of food with family and/or neighbours (Daňek et al. 2022).

The juxtaposition between quietly sustainable gardeners in Eastern Europe and food activists in Western Europe concerns also skill and learning. One could argue that in Europe, learning together in collective food procurement networks responds to a broadly felt societal need for skills, which is met by 'practivists' (Hoff, Gausset, & Lex 2020). However, such discussions should also include the de facto elitism of some forms of ethical consumption, which are at least potentially class-discriminating because of the lack of access to food gardens or food literacy (i.e., knowledge around what to do with fresh food) for less privileged groups in society. Thus, the contradictions of apprenticeship (Herzfeld 2004) pertain not only to the transformation of learning and knowledge in traditional and modern professional trades (Marchand 2010, 2021) but also to the situated learning that happens in food gardens. Collective food procurement networks can then be equated to forms of 'legitimate peripheral participation' (Lave & Wenger 1991), where the education of attention and resonance with non-humans that are typical of enskillment (Ingold 1993) take place.

The political weight of everyday skills is also underlined in the works of geographer Petr Jehlička. Focusing on East European food self-provisioning in post-socialist countries, he critiques its predominantly economic conceptualisation, and its association with household and informal economy. Instead, Jehlička sees 'quiet sustainability' as a form of commitment that can go a long way 'under the radar' to foster environmental sustainability and food equity, especially in post-socialist countries where sizable portions of the population (43% in the Czech Republic) still practice food gardening as a matter of fact, on a small scale, and in a 'de facto' environmentally friendly way (de Hoop & Jehlička 2017). Instead, where this green infrastructure (straddling practical knowledge and access to land) has not been maintained, re-skilling for sustainable food systems inevitably requires connecting with learning a 'new' craft, such as food preservation or seed-saving. Defining craft is, in turn, complex and controversial. It signifies disparate forms of action that are only loosely connected by the notion of making something by hand, and I suggest that in order to understand the broad range of re-skilling needed in food 'practivism', one needs to connect it with craft learning.

There is a growing literature on the rediscovery of craft (DeNicola 2024; DeNicola & Wilkinson-Weber 2016; Kuijpers 2019; Templer 2020; van Wilgenburg 2024). In *The New Politics of the Handmade*, Anthea Black and Nicole Burisch note how the role of craft has transformed under late capitalism: 'craft in a global context is now mobile, flexible, available on-demand, highly desirable and ready-to-use' (2021, p. 1). In late capitalism, the global circulation of craft artefacts such as hand-made garments, pottery, or craft beer has transformed craft itself. In fact, in this market hegemony, 'craftivism' may obscure deeper 'ethical, environmental and labour issues in the chain of production' (2021, p. 2), thus performing 'craftwashing' (namely, akin to greenwashing, the moral justification of immoral ways of producing and consuming goods). However, craftivism is affirming itself as a form of re-skilling (for example, through knitting, sowing, or embroidery) that also voices forms of 'gentle protest' (Fitzpatrick 2018).

I propose to connect the notions of 'quiet sustainability' and 'practivism' by focusing on how one learns socially through food. What is learnt is firstly to see value in everyday food, secondly, to 'craft' (grow, preserve, process, and prepare) it, and thirdly, to maintain and (re)produce the environment needed to sustain crops (including soil life and biodiversity, but also relations with helpers, neighbours and colleagues). Learning food skills is thus a holistic practice within communities of food. It congeals around the idea and practice of crafting: doing something well, by ourselves, and curating and producing it with certain ideas of beauty and propriety (Martin 2020; Calhoun et al. 2013). Both in the *Food Citizens?* project and in my ongoing fieldwork this focus on learning in communities of practice includes the political economy around them. In the following section, I first build on my ethnographic fieldwork with horticulturalists in food allotments in the Netherlands, to then connect it to the results of the *Food Citizens?* research.

Slow practivism in Dutch urban gardens

Volunteering often features in the Dutch discourse of 'active citizenship', which may take many forms: ranging from participating in a local church or faith group, to volunteering in various charities (*stichtingen*) and NGOs (for example, in environmental conservation, or against food waste, or energy-poverty). This social practice of active citizenship may take the form of workdays (*klusjesdagen*) during which one works (usually manually) collectively. Groups of volunteers perform an array of tasks: pruning trees, cleaning up canal banks, providing meals, or performing citizen science such as bird-counting or insect-spotting to monitor biodiversity in urban areas or in natural parks. The Dutch royal family plays a prominent role in promoting this discourse and practice of active citizenship by participating in initiatives and events on the national volunteering day (*NL doet*: 'The Netherlands does') and through the national *Oranje Fonds* (Orange Fund). In particular, many NGOs, charitable foundations, and start-ups are involved in food gardens. Municipalities foster social cohesion, among other measures, by engaging mental-health patients in 'green services' and including workers with disadvantaged access to the job market in food gardens. In this ethnographic section, I complicate this discourse of participation to reveal the uneasy coexistence of volunteers' deliberations and neoliberal management practices.

Since 2020, I have been volunteering at a historical farm on the outskirts of a Dutch city. While the site of the farm has existed since the Middle Ages, it is now a 'city farm' (*stadsboerderij*), occupying an extensive patch of land within the municipality that is well integrated into the urban settlement grid and transport system. The farm preserves elements of natural habitats (such as meadows for pasturing cows, fallow fields, and perennial shrubs and trees) and wildlife (especially birds and insects). At the time of this research, a group of volunteers was cultivating the land, collaborating with employees who are in charge of soil maintenance, including ploughing and tilling once a year and cutting the grass at the edges of the cultivated plot. While these employees are salaried (although most of them work part-time), volunteers have been working for the last 15 years in exchange for free access to small, self-managed patches of land (not exceeding 3 by 4 metres) for their own use. They gave over the majority of the harvest they gathered to the farm shop, which operates regular hours six days a week, and organised stalls selling fresh organic veggies grown on site on the weekends. Some of the vegetable patches were kept for school children, who visit regularly as part of classes taught by a farm employee and a volunteer.

Both the school and the café contribute to developing connections between the residents of the municipality and the farm. However, the municipality recently stressed that they would like to see such involvement with residents increased, in order to serve the entire community. That said, the city farm does engage the public, especially young families, through the extended hours of the shop and the café and a playground near a pond. The farm also provides covered space for assemblies, workshops, and small events. Fermenting and composting workshops are regularly held in the same building as the café. The building also hosts employee offices, a kitchen suitable for collective meals, toilets, and a seminar room. Barns housing sheep and cows, rabbits and peacocks are open to the public on weekends. Their refuse is used as manure in the fields and the animals graze in the meadows.

The coexistence of the farm and its garden allotments is not straightforward. For example, keeping animals captive is increasingly contrary to animal welfare concerns, which some of the allotment gardeners feel strongly about. However, food production and social participation go hand-in-hand with a discourse of sustainability. By this I mean literally the language, deliberations, and debates individuals engage in while practicing food gardening. Such discussions include, for example, how to get rid of slugs in an extraordinarily wet season or how to replenish the soil's productivity that has been lost to nitrogen depletion.

Figure 10.1 captures a moment during the harvest festival of September 2024, which was actually rather subdued. Firstly, there wasn't much food at all. It had been a disastrous growing season: the crops that didn't go mouldy because of excessive rain and lack of sun were eaten by slugs. This followed an excessively dry year in 2023 that had exhausted everyone who watered the crops in the hottest months. Secondly, our mood was sombre because of a recent announcement by the municipality. In light of national budget cuts, the municipality was cutting the funds that support the NGO managing the city farm by 10%. At both levels, climate change and the call for yet another round of austerity in public spending, we perceived an attack on our own sustenance as a group.

FIGURE 10.1 Potato harvest on the city farm, September 2024 (photo by author)

Most city farms in Dutch cities are not individual grassroots projects; they are organised as foundations or NGOs (like this one) that manage several locations. Some of these have a more recreational purpose; others receive extra financial support from the municipality's health services in order to support mental health patients. 'Our' organisation manages a network of about a dozen city farms and has directly inherited what once was a green service management department of the municipality. More than ten years ago, this task was transferred to a foundation, which employs the staff, but remains for all intents and purposes dependent on the political will and economic support of the city government.

The budget cuts galvanised volunteers and employees to protest the municipality's 2024 austerity decision. The volunteers clearly understand the direct connection between the survival of the NGO and the management of the farm and its food garden. I, too, became involved by leveraging a little bit of symbolic power as an academic, by writing a letter to the City Council explaining that the gardens of this organisation receive visits from all over the world and are also cited in academic literature. In addition, there was a call to action, petitions were signed, and a general mobilisation of the community occurred to protest the budget cut. In July 2024, the farm's employees and volunteers demonstrated together outside and inside the Town Hall (see Figures 10.2 and 10.3). With this mobilisation and the petitions and

FIGURE 10.2 Demonstration inside City Hall, July 2024: Handing over the petition protesting budget cuts (photo by author)

FIGURE 10.3 Demonstration outside the City Hall to protest the budget cuts, July 2024 (photo by author)

letters to the municipality, the protesters succeeded in keeping the cuts below 200,000 euros, but still...

While the cuts were postponed and reduced for a couple of years, the protest immediately triggered a discussion of plans for the near future (2026!), which were emphatically cast in a market-driven framework and language. In its newsletter, the NGO managing the city farm announced that we would have to become 'structurally stronger' and adopt a 'new management structure'. We would also have to expand our financial support 'from more sources' and strengthen a 'binding role' with enhanced visibility within the city, including more 'activities and events' for a broad range of residents.

In sum, the volunteers' own mobilisation to respond to the municipality's demands resulted in a plan for large-scale innovation and restructuring of their own participation and a comprehensive reorganisation of the food garden. The invitation to the NGO to render themselves structurally stronger with a new management structure and to find more sources of funding translated into a plan to establish a 'community supported agriculture' project (CSA). According to a presentation made to the group of volunteers, this project would involve intensifying the cultivation on the entire grounds of the farm, a total of 0.35 hectares (ha), which would allow the farm to produce 100 to 140 'shares' of the harvest to be sold to the public in veggie boxes. According to this presentation, this could even project to a potential scaled-up scenario involving 0.7 to 0.9 ha in food production with up to 360 subscriptions by members of the public signing up to buy veggie boxes before the season. This would allow the city farm to 'be more visible' and 'serve the entire community', employing 1.5 to 2 full-time farmers.

The ethnographic turning point in this process was marked by a couple of meetings with the volunteers, which happened during the winter off-season. Pep talks were delivered to us by a hired consultant in order to 'activate' us within four months (from December to March) into becoming part of a 'movement of movements'. What we feared was that a significant increase in the scale and pace of the gardening routine was coming, which would, in effect, transform slow practivism into an (unpaid) job and possibly undo the everyday sustainability politics that had kept us engaged for months and years. We had been resisting slugs without using pesticides, dealing with failing crops, and experimenting with mulching and 'no-dig' techniques. The most experienced gardeners experimented with diversifying crops to place less pressure on a plot of land already nitrogen-depleted by years of cultivation. The new plans alerted us with a jolt to the uneasy coexistence between diverse visions of what it means to practice 'sustainability' in an urban setting.

Through the grapevine, namely via our programme manager and in sporadic meetings with the NGO managing the garden, we came to understand that our access to the garden was considered the most controversial aspect of our involvement in the farm. How could we be a garden for the community if single individuals have the privilege of cultivating public land for free instead of paying rent? How could this privilege be explained? Were we not a clique, a niche circle that is ultimately individualistic? Should we not contribute in a more disciplined way to the growth of more clearly labelled forms of food activism such as regenerative agriculture, ecosystem restoration, short food supply chains, community-supported agriculture, bio-based solutions for systems change? Wasn't the fact that we saw ourselves as just gardeners rather than part of a movement proof of our ultimately apolitical and therefore egotistic engagement? With the new plans, in exchange for volunteering in the new 'community garden', we could pay like everyone else for a veggie box, instead of cultivating our 'own piece' (*eigen stukje*) of the garden. Then, our volunteering would be truly free and disinterested and would not detract from other people's access to the fruits of the garden. We were presented with concepts such as 'inspiration' and 'activation', which should help us to deliver 'co-creation' and 'execution' of the CSA plan. In sum, we found ourselves called 'food heroes' in a call to arms to establish a veggie-box scheme.

This is an example of how craft practice occupies an ambivalent space between hegemonic ideas of future-making and the patient, slow re-skilling through 'dull' timescales and apparently 'apolitical' tasks. The former is characterised by the exciting discourse and practice of 'projects' carrying slogans and accelerated timelines that pertain to neoliberal 'activation' of citizens, by way of colourful PowerPoint slides that jolt communities and nature into production. The latter is made of the daily rounds of watering vegetables when it's too dry and the morning and evening rounds of picking up slugs from wet plots when it's too rainy. What worries some of these 'slow' food gardeners is that intensifying the rhythms and scales of production even on a rather extensive city farm could decrease the wealth of biodiversity found in our loosely cultivated plots that teem with butterflies, insects, connect *micelia* (fungi), and host wild species of plants and flying beasts (Bentvelzen 2019).

In the face of these diverse challenges, defining an appropriate scale and pace for food gardening as green practivism is tricky. The materiality of field, tools, indoor spaces, and rhythms of collective work plays an important part in including a healthy balance of rediscovering and celebrating taste and artisanship, taking pleasure in food harvesting, preparation, and conviviality. However, groups like ours do not impact sustainability politics in any measurable way. Market and global (food) economies remain the orthodox frame of reference for the management of the garden's organisation and for the institutions funding the slow practices cultivated in our garden.

The skilled practices of focusing on taste, artisanship, and pleasure in food preparation (for example, making sambal, a piquant sauce made out of chili peppers that are grown on site) have also been thematised and embraced by movements such as Slow Food and *Cittaslow*, which have also been charged with being elitist and non-inclusive because of the exclusivity of the prices they leverage for heritage crops, niche recipes, and singular culinary productions (Siniscalchi 2023; West & Domingos 2012). While fair, such critiques are sometimes not constructive. That these skills can be found in traditional farming practices and are now becoming residual even in urban food gardening tells us how they are actively being made redundant to the market economy and the global food system. Initiatives such as Slow Food's Ark of Taste (a project aimed at visually and sensorially archiving foodstuffs and tastes under threat of cultural extinction), for example, try to synergise with governance in the interest of 'traditional knowledge' (TK) and 'traditional ecological knowledge' (TEK), i.e., articulating concepts and practices of autochthony and conviviality while documenting food crafting skills (see Fontefrancesco et al. 2024).

Cultivating slow skills, therefore, can be a form of quiet practivism that subtly resists the market hegemony of homogenising sense-scapes (La Fleur 2023), through a re-skilling of the senses against global extractivism (Chagnon et al. 2022) and by cultivating a subversively slow timescale (Forno & Wahlen 2022). What motivates most of the volunteers I have interviewed during my research is neither abstract principles of solidarity and diversity, nor equally abstract notions of food transition. What motivates them is engaging with nature and producing food to gain satisfaction, cultivate community, and to experiment and learn at one's own pace. In other words, our experience of learning through volunteering is not positioned in grand innovation narratives. In the next section, I contextualize these conclusions about what motivates volunteers in the framework of the *Food Citizens?* research across Europe.

Slow practivism in other European cities

In parallel with my fieldwork, the other *Food Citizens?* project members also conducted ethnographic research on the juxtaposition of gentrification, professionalisation, and the very different ideas of re-skilling found in European cities. In the context of gardening practices with diverse local histories, different timescales, and traditions of 'quiet sustainability', the *Food Citizens?* team explored similar uneasy coexistences in the aforementioned post-industrial cities of Gdańsk, Rotterdam, and Turin (Grasseni et al. 2022).

The assumption leading the *Food Citizens?* project was that by re-introducing the collective, i.e., the social element in the analysis of food provisioning, we would be investigating not only styles of food procurement but also societal styles of participation and learning. While one can critique what 'alternative' procurement actually means (Fuller et al. 2010), the language of collective food procurement can refocus us on people's practical ways of participating in society: putting people back in the picture means observing their cultural styles of procurement; it also means focusing more precisely on their *collective* styles of procurement, i.e., on how they socialise the act of procuring food.

Practices of food procurement exceed techno-scientific imaginaries about machine- and science-driven global food systems, and one-size-fits-all 'sustainability fixes'. All across Europe, one can find a rich diversity of histories, styles of governance, economic standards, societal debates, ways of getting by, and modes of participation. From the realisation of this diversity began an agenda for ethnographic comparison. Among the methods we used was the production of an 'i-doc', an interactive digital platform that embeds a multiple-author documentary. It makes explicit connections between case studies within and across regions, for example, by categorising them under the topic of skill (De Musso et al. 2022).

One of the insights of the *Food Citizens?* project was that, unsurprisingly, learning food procurement skills in a collective dimension is not an ideational process but a grounded, contextual, relational process. Ideational and ideological work gets translated and inscribed in very localised contexts which connect people with specific materials, ecosystems, networks of neighbours, and governance agendas. These agendas are broad, sometimes contrasting, and grounded in local political processes (Grasseni et al. 2022).

Collective forms of food procurement, especially in the guise of self-provisioning, shift the focus away from the understanding of activism as a form of ethical consumerism—that is, as a form of individualised, rational choice within the hegemonic discourse of global consumerism that dominates the visual imagery surrounding the future of food. If we look at 'future foods' in the global food system today, we see, on the one hand, the global seed industry with its gargantuan capacity for research and development proposing ever more technology (e.g., the use of sensors and geolocation, hydroponic architecture, and greenhouses to allegedly decrease the use of fertilizers and herbicides). On the other hand, the global imaginary of the individual consumer in the supermarket dominates, making 'choices' and leveraging price, quality, and taste as well as morality, 'fairness', and food miles (a commonly accepted measure of the fuel used to transport food).

There is very little space in this global techno-societal imagery for nuance, for the local context, and for the sociocultural dimension, namely, for agency, conflict, and diversity. There is also very little space in this imagery for *relationships* among humans and between humans and non-humans. The underlying expectation is that sustainability fixes will be found for all problems, such as those relating to food access. This inevitably replicates and increases the insertion of technology into daily practices. Such practices range from looking up food provenance and quality labels on smart phones through QR codes, to adopting the recommendations of apps against 'food waste' by buying mass-produced food that is 'too good to go' (the name of a popular app through which Dutch consumers can visit shops at a certain late hour to buy up leftover food at lower prices).

The *Food Citizens?* project combined and contrasted grassroots perspectives through analytical categories. 'Solidarity', 'diversity', 'skill', and 'scale' were our chosen entry points, i.e., the angles or viewpoints from which to ask questions. For example, questions to ask included: Which skills do people involved in collective food procurement acquire or lack? How do they operate across and within diverse communities? Do their networks scale 'up' or 'out', and how? How do they interpret and articulate solidarity? Working with a team of three PhD students, two research assistants, and two post-doctoral researchers for over five years, the project incorporated audio-visual training into the PhD trajectory so that we could not only narrate our ethnography with words, but also visualise our fieldwork. The project's i-doc allows the user to digitally 'navigate' 50 case studies in the cities of Gdańsk, Rotterdam, and Turin, to juxtapose and contrast different kinds of communities of food and how they relate and compare with each other (De Musso et al. 2022).

Gentrification as an urban dynamic is common to all three sites. For example, in Poland, a tradition of 'food self-provisioning' built around informal networks of exchange of food preserves from the food allotments previously assigned to factory workers is associated with an aging generation. On the other hand, as investigated by Ola Gracjasz, fermenting workshops are being taught to professional working women living in the city, who have lost touch with this family tradition but want to reinvent it on their own terms (Gracjasz 2024).

As per her research, family allotments are key to handing down gardening and processing knowledge, because allotment gardeners work the land and make preserves. In Poland, 'family allotment gardens', *rodzinne ogrody działkowe* (ROD), continue to be managed by the Polish association of allotment gardeners, PZD. Through its nationwide organisation, PZD manages all urban allotments in Poland and serves as a spokesperson for allotment gardeners in municipal or government negotiations. Some allotments have long histories and have been cultivated over generations. For example, right after World War II, plots on the hill overlooking the port of Gdańsk were allocated to the workers of the shipyard. Plots usually include small summer houses and a communal building for holding events. They are embedded in the fabric of local neighbourhoods and have relationships to nearby schools, too.

But 'new' or 'alternative' vegan places in the city also play a role in reconnecting with this skill, for example, organizing workshops and meetings for people interested in becoming new members of a ROD. Re-learning the old-time skill of preserving food in jars comes up not only in fancy gentrified quarters but is also taking root in neighbourhoods. One example is Falowiec, a vegan café located in the homonymous neighbourhood of Gdańsk in Europe's largest residential building constructed during socialist times. Here, during the COVID pandemic, making food jars for online sales became a new version of an old skill (Gracjasz 2024).

The gentrification of the European urban green is also a daily reality in Rotterdam, where grassroots social gardens often go hand-in-hand with the feeling of empowerment and autonomy that comes with new skills and knowledge. But these green projects are happening next door to urban mega-developments that make the future of urban gardens very precarious. For example, consider *Moestuincomplex Gordelweg*, an allotment gardens complex located on a green strip between the inner and outer beltway of north Rotterdam and studied by Vincent Walstra. It houses school gardens and over a hundred individual vegetable plots. It includes animal-friendly spots like birdhouses, a pond, and beehives. The board of this complex organises skills workshops for the gardeners, for example, to educate them on how to compost. In one of the videos of the i-doc mentioned above, we are told that 'kids see Romanesco broccoli as art pieces and don't want to harvest them' (Walstra 2022). Community gardens often occupy liminal spaces at the margins of housing estates or near infrastructures such as railways or roads, which raises concerns about how long these gardens are going to last, and what they can do long-term to educate their users in rooted sustainability practices.

In Turin, the proliferation of greening policies also raises issues about who decides which skills and which aesthetic should be predominant on municipal land that is managed by third-sector NGOs. The messy, ad hoc food gardens of ex-Fiat factory workers, for instance, had to give way to projects run by landscape architects for which residents are expected to volunteer, but on someone else's terms. Anthropologist Maria Vasile investigated an urban regeneration project in Turin, where a community garden was set up thanks to the participation of local residents with experience in gardening. The goal was to make more orderly, more regulated, and more coordinated use of public soil as mandated by the municipality. But the new project wiped out existing gardens that didn't have an operating license and looked more liminal, less efficient, and certainly less orderly. With their unruly recycling practices, these 'illegal gardeners' went against the grain of gentrified expectations of what a community garden should look like. Instead of beautifying a public green, they confirmed expectations of degradation in peripheral areas, because, for example, they reused and re-adapted building materials from skips or old bedframes from rubbish dumps (Vasile & Grasseni 2022).

In sum, the 50 cases described in the foodcitizens.eu website clarify several aspects common to them all. First, gentrification affects all three cities. Second, one finds a very diverse range of practices and engagements, some of which are quieter forms of activism. Third, the relationship between food activists and institutions is complicated: it varies from one of reciprocal suspicion and disregard (in Italy and in some Polish cases) to one of institutionalised participation and mutual expectations of collaboration (in some Polish cases and in the Netherlands). These case studies articulate the contradictions mentioned in the previous section: the expectation of volunteers to be active citizens but also the underlying tension between volunteerism and collective agency. My ethnographic case resonates with these findings because it is precisely the forces of gentrification that make free access to municipal land difficult to justify in neoliberal terms. The gardeners' resistance, i.e., their slow but long-term craftsmanship, vis-à-vis top-down mobilisation and management from above, illustrates the contrast between slow practivism and productivity, marketisation, and formalisation. These forms of slow resistance emerge in unexpected ways, for example, by the resistance of gardeners to being 'activated' according to timescales that do not belong in their slow practice.

Conclusion

This chapter highlights the social architecture of green communities with a case study about the interaction between an institution providing access to public land through volunteering (in this case, an NGO) and a community of practice involved in urban gardening. It outlines the 'practivism' of the garden volunteers as a form of slow resistance against hegemonic expectations, for example, about active citizenship. The recent budget cuts by the municipality heavily influenced the decisions of the NGO managing the garden, which is mandated to provide green services to the resident population through city farms.

I analysed this case first through an ethnographic narrative and then in comparison with the case studies presented in the *Food Citizens?* i-doc. I also studied the case against the backdrop of a literature review on skill, craft, and collective learning. My ethnography and the i-doc help us focus on issues of gentrification and neoliberal governance, which have also emerged in other food gardens in the Netherlands, Italy, and Poland. While forms of grassroots engagement vary, they show the many ways in which the relationship between slow practivists and government institutions can be complicated and potentially conflictual.

Volunteering is expected and encouraged in the Netherlands as part of an ideal imagery of 'active citizenship', but this can also be an excuse for city and national government to withdraw more and more from providing services to citizens, including 'green services' such as running a city farm. In this case, the activation of residents in the name of active citizenship can be a hegemonic construct that enrols unpaid volunteers to provide well-being services and environmental stewardship in their spare time, in exchange for being granted access to public land and its recreational benefits.

However, food procurement is about more than producing food. It is about relating to each other and to (urban) nature through food. Even on city farms, sustainable food production is local, is slow, is hand-made, is crafted, and is often more expensive than large-scale production. Therefore, it should not be quantified in terms of output or even of number of citizens served by veggie boxes that are produced on site. The value created on site exceeds that of food quantity or money saved at the supermarket.

The *Food Citizens?* i-doc showcases diverse imaginaries of who might be the protagonists of food transition. This is an open question. Should urban food gardeners be seen as food citizens at all? What kind of people are they? Is it going to be the young, educated, alternative, digitally literate, or the ageing woman with craft knowledge and a network of relations that are used to pass on her 'quiet sustainability'? Perhaps *both*! Collective food procurement networks in European cities produce a diversity of language and imagery about food sustainability, health, and urban inclusion. However, this diversity is seldom inscribed in policy and has anish eco-villages and social housing estate so far had minimal impact on global food systems' markets and politics.

This case study further supports the need to investigate how learning by experience and direct engagement (in food procurement as well as in craft skills and traditional ecological knowledge more broadly) is important in debates about green communities. Greater understanding of this process would contribute to resisting top-down projects about 'scaling up'. In a crucial turning point of municipal budget cuts, the volunteers I have worked with for years succeeded in communicating their value and bought time to work on the survival of their city farm. This urban gardening group expresses resistance by its perseverance, persistence, and its sheer existence. However, a broader local engagement in green transitions would also require communicating what is gained in this kind of slow practivist engagement rather than thinking in terms of maximising food production at urban farms. Ultimately, this approach to building green communities is a slow form of resistance that is grounded in cultivating the perception of food value in skills, rather than monetising them.

References

Agyeman, J., Bullard, R. D., & Evans, B. (Eds.). (2003). *Just sustainabilities: Development in an unequal world*. MIT Press.

Bentvelzen, T. (2019, 2 May). Deze plantendeskundige brengt elk zeldzaam plantje in Utrecht in kaart [This botanist takes into consideration each and every rare plant in Utrecht]. *Algemeen Dagblad*. www.ad.nl/utrecht/deze-plantendeskundige-brengt-elk-zeldzaam-plantje-in-utrecht-in-kaart~afef607b/

Black, A., & Burisch, N. (Eds.). (2021). *The new politics of the handmade: Craft, art and design*. Bloomsbury.

Calhoun, C., Sennet, R., & Shapira, H. (2013). Poiesis means making. *Public Culture*, 25(2), 195–200.

Chagnon, C. W., Durante, F., Gills, B. K., Hagolani-Albov, S. E., Hokkanen, S., Kangasluoma, S. M. J., Konttinen, H., Kröger, M., LaFleur, W., Ollinaho, O., & Vuola, M. P. S. (2022). From extractivism to global extractivism: The evolution of an organizing concept. *The Journal of Peasant Studies*, 49(4), 760–792. https://doi.org/10.1080/03066150.2022.2069015

Daněk, P., Sovová, L., Jehlička, P., Vávra, J., & Lapka, M. (2022). From coping strategy to hopeful everyday practice: Changing interpretations of food self-provisioning. *Sociologia Ruralis*, 62, 651. https://doi.org/10.1111/soru.12395

De Hoop, E., & Jehlička, P. (2017). Reluctant pioneers in the European periphery? Environmental activism, food consumption and 'growing your own'. *Local Environment*, 22(7), 809–824.

De Musso, F., Gracjasz, O., Grasseni, C., Poulopoulou, M., Smith, R., Stalenhoef, H., Vasile, M., & Walstra, V. (2022). *Food Citizens?* Interactive documentary. https://foodcitizens.eu/idoc/

DeNicola, A., & Wilkinson-Weber, C. (2016). Introduction: Taking stock of craft in anthropology. In C. Wilkinson-Weber & A. DeNicola (Eds.), *Critical craft: Technology, globalization and capitalism* (pp. 1–16). Bloomsbury Academic.

DeNicola, A. (2024). Craft. In H. Callan & S. Coleman (Eds.), *The international encyclopedia of anthropology* (pp. 1–16). Wiley.

Fontefrancesco, M. F., Vandebroek, I., & Pieroni, A. (2024). Editorial: Traditional knowledge in food activism and governance. *Frontiers in Sustainable Food Systems*, 8, 1450833. https://doi.org/10.3389/fsufs.2024.1450833

Forno, F., & Wahlen, M. (2022). Environmental activism in everyday life. In M. Grasso & M. Giugni (Eds.), *The Routledge handbook of environmental movements* (pp. 434–450). Routledge.

Fuller, D., & Jonas, A. E. G. (Eds.). (2010). *Interrogating alterity: Alternative economic and political spaces*. Routledge.

Gausset, Q., & Jensen, P. D. (2024). Living sustainably in a Danish eco-community: How social and physical infrastructures affect carbon footprints. *npj Climate Action*, 3(1), 33. https://doi.org/10.1038/s44168-024-00113-5

Gracjasz, A. (2024). *'In transformation': Trust, participation, and new socialities around collective food procurement networks in Gdańsk* [Doctoral dissertation, Leiden University]. https://hdl.handle.net/1887/4108617

Grasseni, C. (2018). Grassroots-responsible innovation initiatives in short food supply chains. In A. Kalfagianni & K. Skordili (Eds.), *Localising global food: Short food supply chains as responses to agri-food system challenges* (pp. 41–54). Routledge.

Grasseni, C. (2024). Re-territorialising skills? Insights from ethnography on solidarity-economy food activism. *Sustainability Science*, 20(2025), 1639–1651.

Grasseni, C. (2025). More than visual: The apprenticeship of skilled visions. *Ethos*, 53(1), e12372. https://doi.org/10.1111/etho.12372

Grasseni, C., De Musso, F., Gracjasz, O., Smith, R., Vasile, M., & Walstra, V. (2022). Reskilling for sustainability: A perspective from comparative ethnography on collective food procurement. *Kritisk Etnografi* [*Swedish Journal of Anthropology*], 5(1–2), 135. www.diva-portal.org/smash/get/diva2:1717941/FULLTEXT01.pdf

Fitzpatrick, T. (2018). *Craftivism: A manifesto/methodology*. https://talfitzpatrick.com/craftivism-manifesto

Hankins, J. (2023). *Food for good*. Fondazione Giannino Bassetti.

Herzfeld, M. (2004). *The body impolitic: Artisans and artifice in the global hierarchy of value*. University of Chicago Press.

Herzfeld, M. (2007). Deskilling, 'dumbing down,' and the auditing of knowledge in the practical mastery of artisans and academics: An ethnographer's response to a global problem. In M. Harris (Ed.), *Ways of knowing* (pp. 91–112). Berghahn Books.

High, S., MacKinnon, L., & Perchard, A. (Eds.). (2017). *The deindustrialized world: Confronting ruination in postindustrial places*. University of British Columbia Press.

Hoff, J., Gausset, Q., & Westergaard, S. L. (2020). *The role of non-state actors in the green transition*. Routledge.

Ingold, T. (1993). The temporality of the landscape. *World Archaeology*, 25, 152–174.

Ingold, T. (2000). *The perception of the environment: Essays on livelihood, dwelling and skill*. Routledge.

Ingold, T. (2013). *Making: Anthropology, archaeology, art and architecture*. Routledge.

Kuijpers, M. (2019). The future is handmade [Short documentary]. *Craftsmanship Magazine*. https://craftsmanship.net/videos/

LaFleur, W. (2023). Storying sensuous atmospheres of peaches and wheat: Towards new horizons of imagination and possibility for sustainable food systems. *Ethnologia Fennica*, 49(2), 66–100. https://doi.org/10.23991/ef.v49i2.112979

Lave, J., & Wenger, E. (1991). *Situated learning: Legitimate peripheral participation*. Cambridge University Press.

Marchand, T. H. J. (2010). Making knowledge: Explorations of the indissoluble relation between minds, bodies, and environment. *Journal of the Royal Anthropological Institute*, 16(S1), S1–S21.

Marchand, T. H. J. (2021). *The pursuit of pleasurable work: Craftwork in 21st-century England*. Berghahn.

Martin, T. (2020). *Craft learning as perceptual transformation: Getting 'the feel' in the wooden boat workshop*. Palgrave Macmillan.

Sekules, K. (2020). *Mend! A refashioning manual and manifesto*. Penguin.

Siniscalchi, V. (2023). *Slow food: The economy and politics of a global movement*. Bloomsbury.

Smith, J., & Jehlička, P. (2013). Quiet sustainability: Fertile lessons from Europe's productive gardeners. *Journal of Rural Studies*, 32, 148–157.

Stanton, C. (2017). Between pastness and presentism: Public history and local food activism. In J. Gardner & P. Hamilton (Eds.), *The Oxford handbook of public history* (pp. 217–236). Oxford University Press.

Stead, M., & Coulton, P. (2022). A more-than-human right-to-repair. In D. Lockton, S. Lenzi, P. Hekkert, A. Oak, J. Sádaba, & P. Lloyd (Eds.), *Design research society 2022: Bilbao, 25 June–3 July* (pp. 1–21). Design Research Society. https://doi.org/10.21606/drs.2022.718

Templer, A. (2020). Techniques for shifting economic subjectivity: Promoting an assets-based stance with artists and artisans. In J. K. Gibson-Graham & K. Dombroski (Eds.), *The handbook of diverse economies* (pp. 419–427). Edward Elgar.

van Wilgenburg, W. (2024). *Ambacht in Beeld Festival: An initiative of Stichting Ambacht in Beeld* [*Craft in Focus festival: An initiative of the foundation Ambacht in Beeld*]. https://ambachtinbeeldfestival.nl/en/over/

Vasile, M., & Grasseni, C. (2022). Visions of the urban green. *Anthrovision*, 8(1), 1–35. https://doi.org/10.4000/anthrovision.6812

Walstra, V. (2022). Moestuincomplex Gordelweg [*Allotment complex Gordelweg*]. Video. In F. De Musso et al. (Eds.), *Food citizens?* Interactive documentary. https://foodcitizens.eu/idoc/

West, H. G., & Domingos, N. (2012). Gourmandizing poverty food: The Serpa cheese Slow Food presidium. *Journal of Agrarian Change*, 12, 120–143. https://doi.org/10.1111/j.1471-0366.2011.00335.x

11

COMMUNITY GARDENS AND THE ACT OF FOOD SHARING

The impact of community engagement

Susan Andreatta

Introduction

The chapters in this book examine the social architecture of green communities from a grassroots and practivist perspective with the goal of providing a clearer understanding of how alternative social organisations and processes contribute to liveable and sustainable local communities. As I am an applied anthropologist and academic, I have taught courses in environmental anthropology, and in food and culture, and embedded degrowth into syllabi, which enabled me to demonstrate practivism in the classroom. I have also been involved in years-long activism around local agriculture, community gardens, and equitable food access, leading by example in both curricular and co-curricular activities (Andreatta 2025, 2022; Andreatta et al. 2023; Andreatta & Hoskins 2022). Therefore, this chapter is a case study that examines a community garden through the lens of community engagement in order to encourage readers to think about applied local agriculture and food projects that can be organised as class projects at their place of residence and in their communities.

Project background

Over 25 years ago, a few colleagues and I attempted to start a community garden on the campus of the University of North Carolina at Greensboro (UNCG). It went nowhere; it turned out that we were ahead of the sustainability movement. Rather than be dissuaded, I pushed forward and got involved in local agriculture and food systems. These efforts enabled me to start Project Green Leaf (PGL) in 2001 as a community-engaged approach

DOI: 10.4324/9781003769064-12

to assist farmers with connecting to the public and the public connecting to local farmers. In other words, we were building two-way bridges with multiple on-ramps and exits to welcome people to join environmental groups relating to food and culture, sustainability and equitable local food access.

For example, at the turn of the 2000s, small-scale tobacco farmers were transitioning from tobacco into diverse commodities and needed new markets. Also at that time, organic crop production was gaining traction among growers and early 'foodies'. PGL and agricultural extension agents assisted in giving workshops to discuss these local markets. In time, PGL broadened its scope to include local restaurants, primarily those that created their own menus and had control over sourcing fresh produce. Later, independent small-scale coastal fishermen also reached out to PGL in order to make more connections in the local food community and to benefit from local markets. (These farmers required more assistance in direct marketing and in building community around a locally sourced agri-food system.)

From its inception, PGL has been externally and internally funded to support a wide array of projects. As one of our major donors explained, '[They] like to support PGL because the organization helps to teach others how to help themselves'. Other efforts were supported by external grants. This funding helped PGL evolve into supporting field-to-table and sea-to-table activities by working with existing or establishing new local farmers' markets, i.e., community-supported agriculture arrangements (CSAs) (Andreatta 2005; Andreatta, Rhyne, & Dery 2008), and community-supported fishery arrangements (CFAs) (Andreatta & Parlier 2010; Andreatta, Nash, & Martin 2011). These efforts increased the number of farmers' markets in the surrounding area, which paralleled the national increase in the number of farmers' markets; they also increased the number of farmers attending markets and the public supporting them.

A long-term concern has been aging farmers and marketgoers and a declining number of younger people participating in these projects. However, with increasing awareness and accessibility, more small farmers and younger farmers have begun to join the ranks, and young families are also supporting direct marketing and local food systems (United States Department of Agriculture (USDA 2023).

Projects emanating from PGL prompted me, UNCG staff, and faculty to attempt to get university administrators on board with the idea of a campus garden. After nearly a decade and many changes in the university's administration, I provided administrators with students' class projects that outlined how an on-campus community garden would enrich their academic experience. The students tied their vision for the garden to the university's core value of sustainability,[1] which convinced the administrators to support it. (A lesson learned here is the importance of students' voices and how effective their voices can be to influence change at institutions.) Permission was granted in July 2010, and ground was broken for the UNCG Campus Gardens.[2] Plenty of sweat and sweat equity was invested in the garden. And after my participating in the UNCG's Campus Gardens (n. d.) all these years, an enriching story can be told about a community-engaged campus gardens and the resulting growth in students' interest in gardening, food, sharing, and cooking.

The purpose of this chapter is multi-fold. First, I begin by addressing community engagement and its link to practivists and then weave the UNCG Campus Gardens in as an exemplar. Second, I describe the steps involved in establishing an on-campus community garden while illustrating the various ways in which community is created and engaged to support the garden from a practivist's perspective. Lastly, I conclude with the lessons learned from this experience of working in the local food system for nearly three decades and incorporating these experiences into the classroom. The most important goal of this effort is to inspire future generations to connect with their local food systems, to farm, to garden, and to create community around food. The viability and scalability of these efforts is seen in the outcome of this journey.

Community engagement and practivists

Thinking about what community means is a good place to start before jumping into community engagement. For some, community is a term that enables one to define groups, such as those who may share an interest or who are stakeholders and members of citizen groups. Others like to think of community slightly differently, not necessarily as people first, but rather a geographic location, such as a community of place or similar interest. It could be a community of practice or a community of affiliation or identity, all which can be in person, symbolic, or virtual (Community Places 2014; McCloskey et al. 2024). For example, one might identify with a community because of shared allegiance to a sporting club such as a community of football fans.

To motivate a community or a group to work together around an issue or cause or something that brings them together and activates them into action leads us to community engagement. The United States Centers for Disease Control (CDC) defines community engagement as 'the process of working collaboratively with and through groups of people affiliated by geographic proximity, special interest, or similar situations to address issues affecting the well-being of those people' (CDC/ATSDR 1997). The CDC sees community engagement as a means to bring about environmental and behavioural changes to improve the health of a community and its members. Moreover, community engagement 'often involves partnerships and coalitions to help mobilize resources and influence systems, change relationships among partners, and serve as catalysts for changing policies, programs, and practices' (CDC n.d.).

This definition is an appropriate one for this chapter as it captures the efforts and activities at both PGL and the UNCG Campus Gardens.

Because there are many forms of community and explanations of how individuals or groups align or identify with a community or place, it behoves practitioners to learn how individuals connect and enter into relationships that form shared communities and motivate them. One way in which this can be accomplished is through cultural awareness. Donna McClosky and her colleagues note that:

> culture shapes identities and fosters notions of community, and it shapes how individuals and groups relate to each other, how meaning is created and how power is defined. Further, culture shapes ideas about partnerships, trust, and negotiation. Therefore, culture shapes the process of community engagement, and effective engagement requires an understanding of culture.
>
> *(McClosky et al. 2024, 2)*

To build and maintain community relationships, practitioners should also understand the cultural dynamics of specific groups and/or institutions in order to effectively collaborate, build respect, and gain trust. At its most basic level, community engagement, also referred to as civic engagement or public participation, endeavours to improve the quality of life while making a positive difference in one's community that manifests itself through action, knowledge, skills, values, and motivation (McClosky et al. 2024).

Empowering community members to self-actualise leads to engaged community members, also known as practivists. Practivism is defined as 'a form of activism that is based on changing one's own practices rather than changing those of others' (Hoff et al. 2020). As a concept, practivism emphasises the importance of engaging in activities that support social and environmental causes in a more locally and inwardly oriented professional or personal manner (Hoff et al. 2020). A practivist embraces personal growth, practical skills, sustainability, energy efficiency, and community engagement in their daily lives. They use their lifestyle choices, purchasing decisions, practical skills, and actions to positively impact the earth and other communities (Postila & Erkisson 2024). For example, a practivist might run a successful business to do a social good, such as refashioning and repurposing thrifted clothing for those on a limited budget. Critical to the success of a practivist, however, is communicating well and in an adaptable manner to motivate action for positive change.

There are numerous ways in which a practivist can participate in community engagement. For example, there is skill-based volunteering, where professionals might offer their expertise for free, such as IT computer wizards teaching local adults the new technology. There are also project-based initiatives, where practivists work on action-oriented projects that directly benefit the community. Examples here might include maintaining community gardens that would help promote environmental stewardship, food sharing, and food security. Others could include river and verge clean-ups or beautifying vacant lots. Even setting up local farmers' markets to connect farmers and the public to a local agriculture and food system fits in with practivists involved in community engagement. Suffice it to say, practivists use their knowledge to educate and empower community members with a shared mission. And lastly, there is civic participation, in which practivism encourages active involvement in local governance and decision making. Participating in local council meetings or school boards are instances where practivists can advocate for community needs and represent underserved populations.

What becomes clear is that practivists work collectively to address community challenges. They may create support groups for specific community needs by engaging in activist strategies to foster community engagement and create positive social change through direct, skill-based action

I turn now to the University of North Carolina's Campus Gardens as a prime example of the social architecture of green communities from a grassroots and practivist perspective.

The community garden: An exemplar for community engagement

For decades college campuses around the US have turned to campus community gardens to teach about sustainability and concepts related to food justice (Blane et al. 2024; March et al. 2020; Meyer 2022; Murtagh 2024; Yale 2022). On-campus community gardens provide service-learning opportunities and serve as a pedagogical tool for educating students about social justice, civic engagement, community engagement, and personal responsibility for positive social and environmental change.

At the root of studies that have identified communities as 'food deserts,' 'food insecure', or some other label reflecting limited, inequitable access to fresh produce, is a community of poverty. The affordability of fresh produce and the distance one needs to travel to obtain fresh products go hand-in-hand with food insecurity. Those looking for solutions to food deserts or food insecurity often suggest community gardens. However, experience informs us that community gardens are not the complete answer to a community's food insecurity problems (Andreatta 2015).

Nevertheless, other outcomes may be gleaned from a community garden experience beyond providing alternative food choices and fresh produce. For instance, community gardens foster community engagement, and when linked to practivism also foster a sense of connection and responsibility, personal development, and self-care. Practivists at community gardens apply their skills and learn new skills to contribute positively to their society and community while also focusing on their own well-being and mindfulness (Bundy 2024; Lawrence 2022).

Community gardens provide hands-on learning of how to grow produce, herbs, and flowers. The shared gardens on university or school campuses, be they in rural or urban settings, provide camaraderie, skills learning, and produce a myriad of positive environmental, social, and personal opportunities and experiences for those with or without gardening experience. Working in community in these shared gardening experiences provides opportunities to develop one's gardening skills, leadership skills, and life skills. On a campus, such community engagement enables participants to connect with the garden, with other members, and create community with off-campus members, including the public and donors. A number of examples are woven into this chapter to illustrate the various forms and outcomes of community engagement that such gardens express.

Establishing UNCG's on-campus community gardens

The UNCG is an urban campus of 18,000 students located in the City of Greensboro, North Carolina, which has a population of over 250,000 residents. Its campus community garden, to be identified as the UNCG Gardens and the Gardens going forward, was established in July 2010 on property owned by the university. Many people, including the then Chancellor, Provost, Vice-Chancellor of Business Affairs, Associate Director of Administrative Operations, grounds personnel, faculty, and staff were involved in establishing the garden, and some continue to be involved in its use and management.

The location for the garden was selected from several open spaces on campus. Choosing the site was accomplished in consultation with campus ground caretakers, who recognised the need for a water supply, access to electricity, and proximity to residence halls and classrooms. Groundskeepers showed me where they, too, had established tomato plants and sunflowers in open patches near their workshop. In the end, the campus garden site selected was a former house lot measuring 50 ft x 150 ft. (15.24 m x 45.72 m).

Campus involvement has contributed to the success of the UNCG Gardens. For example, students and faculty in the Department of Interior Architecture (IARC) designed the raised beds, garden layout, and benches for an outdoor classroom. IARC students found sources for recycled construction materials for the garden structures. The initial raised beds were constructed from reclaimed oak barn wood, and the frames were lined with cardboard gathered from university facilities (student residence halls and the campus bookstore) and filled with soil, mulch, and compost donated by the City of Greensboro.

Initially, 15 4 ft x 8 ft (1.22 m x 2.44 m) raised beds were assembled and laid out in rows.

Framing out of the garden layout and the initial build was a collective effort by the newly established student garden club, IARC students, staff members, and faculty. For about a decade, woodchips were placed in the rows between the raised beds for about a decade. The woodchips were provided by a firm the university contracted with for tree maintenance and served as a way for the university to reduce waste. These raised bed frames lasted about five years. Subsequent frames have been built from untreated pine, which also lasts about four to five years.

Three months after the house lot was chosen as the garden location, a second house adjacent to the garden was razed and an additional 15 raised beds were arranged. Today, the garden is a community space of 56 raised beds, a teaching station/table, and six benches that serve for convivium and resting. In 2024, a storage shed was erected (12 ft x 12 ft; 3.66 m x 3.66 m). For most of the 15 years, the garden tools were stored at my house, and a few shovels, hoes, and sundry other garden items were left in storage containers (3 ft x 3ft x 5 ft) (.91 m x .91 m x 1.52 m) for community gardeners' use at their convenience. In 2025, solar panels and battery chargers were wired to charge battery-operated tools.

These changes, i.e., where tools are stored in place and recharged, are a way to empower the student garden club to take responsibility for the tools needed to garden and maintain clean pathways between the raised beds. Over the years, when we shifted from corded electric to battery-powered mowers and weedwhackers ('strimmers'), I was the one who charged the batteries during the growing season.

In the early years, we would attract nearly 80 students for a UNCG Gardens workday, which satisfied the service-learning hours that students needed for course credits and also satisfied the students' desire to be outdoors and meet others with shared interests. On these garden workdays, students would lay cardboard between the raised beds and cover the paths with woodchips, which created grassless pathways that did not need to be mowed. A lesson learned, however, was to limit how many students work per hour to avoid not having enough tools or tasks to be completed. Workdays became capped at 30 volunteers an hour and more work was completed in a timely manner, which was critical if rain was forecast.

This system worked well for about ten years, but then the grass began to get ahead of us by taking over the rows. While having the grass take over helped with soil erosion, the pathways between the beds needed to be mowed weekly during the heavy growing season. A past garden club president who learned of our plight on social media donated a battery-operated mower, which solved the increasing maintenance problem. (We also purchased a second manual-operated push mower later on.) Having two people mow and two people weedwhack (or 'strim') also fosters camaraderie for those students doing these tasks. Also, rotating more students into these tasks enables them to be involved with maintenance and volunteer as one of the work options to be completed. It also builds self-confidence and teaches maintenance skills, which some students might not have experienced (if they never had to mow a lawn or grew up without a yard that required care).

In fact, one student recounted with a smile that after she told her mom she found weedwhacking relaxing, her mother then purchased a new weedwhacker for her to use at their home while she was home for the summer. In addition, some past volunteers stay connected to the campus gardens and donate needed items and money after their student days are over. Over the years, we have received gardening gloves, tools (shovels, rakes, hand trowels, and watering cans), seeds, and transplants (vegetables and flowers).

The UNCG Gardens is an organic garden. Soils are deep and rich and attract a nice supply of worms and garden snakes. There is an abundance of beneficial pollinators and insects along with a few not-so-nice ones, and an increasing number of critters who enjoy eating fresh produce (squirrels, birds, groundhogs, and chipmunks) and also cause crop loss at times. (For instance, students have photographed groundhogs eating vegetables in a raised bed.) There is always something to observe and learn that also becomes a talking point among the volunteers. It can begin with 'Is this a weed?' or 'Is this a good bug or should I squish it?'

When the UNCG Gardens was approved for development, university administrators let it be known that there was no funding to support it. Over the years, we have had in-kind university support, but regular funding is not on the cards. This translates into no paid student positions during the academic year, no course release for faculty for coordinating garden activities, and no compensation for staff time. We have managed creatively. Half of the raised beds are cared for by the Student Garden Club, which operates for nine months of the year (weather-permitting) and is ideal for hands-on learning opportunities. The remaining beds are rented to individual gardeners for $10 a semester or $20 for the year. Beds are free to faculty who embed a project involving at least one bed into their syllabi. The summer months in the gardens are the hardest to staff because most of the students and faculty leave the campus. Therefore, in the summer months (May through August) the UNCG Gardens is co-managed by me, two paid student garden workers, and by those renting the beds.

Over the years, donations and grants have provided funds for supplies and summer student workers. One of our largest and most regular donors is the famous musical artist Bruce Springsteen and the E-Street Band. When he and the band come to any city for a concert, they raise funds for local food organisations and food banks. PGL has been a recipient of Springsteen's donations on three occasions.

The limited rent collected does not cover maintenance or annual planting needs. The university provides the space, water, and electricity. Renters provide their own seed or transplants and must water their own beds from the university's source. PGL covers the soil and compost costs, which is available for all the beds. PGL purchases the transplants and seeds for the Student Garden Club and the surplus is made available to those renting beds. Transplants are purchased from local farmers; some are donated by these farmers. Seeds are purchased from an heirloom seed company, Baker's Creek Heirloom Seed Company (n.d.). PGL has also purchased the tools needed for the Student Garden Club, communal workdays, and personal use.

Creating community

Creating community at UNCG Gardens has been a journey. Operating the UNCG Gardens for past 15 years has required teamwork by faculty, staff, students, and support from off-campus partners. The land used by the Gardens belongs to the university and is slated to become transformed into a building and parking lot. However, there has been no move by the university to move these projects along. For now, UNCG Gardens helps maintain the campus sustainability rating. Individuals who manage the physical aspect of the university, be they from recycling, grounds, or maintenance, are supportive of the Gardens. However, this support does not mean they value the garden in the same way as the gardeners do.

Students self-select to volunteer at the Gardens, as do faculty and staff. The key to maintaining this space has been our following a philosophy of leading by example with equity and inclusiveness while embracing practices of sustainability and principles of organic farming. Projects organised for the Gardens foster team building and the development of leadership skills, especially for the students. The garden space has brought staff, students, and faculty together to work on communal work days or to care for the space in smaller groups. There is always work to be done, be it planting, weeding, watering, harvesting, general bed and path maintenance, bug and plant identification, and whatever else may come our way. Students also volunteer at events to spread the word about the Gardens, recruit new members, and maintain social media platforms to connect with interested people, fellow students, and alumni.

Faculty have held classes under the trees where the benches are located to discuss nature, sustainability, and poetry. Other faculty and staff have grown vegetables in beds for their home consumption, while still others have grown flowers to make dye or paper. The biology faculty had their students use the garden for bug and plant identification and for collections for their assignments. The Gardens has been part of several campus nature scavenger hunts. It has become a popular place to gather for some circles, which has led the UNCG Gardens to be listed on the campus map and included as one of the stops on the Spartan Tours for prospective students and their parents when they visit the campus. For some students, knowing that the UNCG has a garden where they can volunteer is a reason for attending this university.

We encourage creative designs and fun as we work together on garden tasks. We share recipes, language, and culture, given the diversity of the volunteers. A faculty member from the Classical Studies Department incorporated the garden into her syllabus for the class and the students used their raised bed to grow a herb garden and labelled their plants in Latin. Students also made a miniature statue for their raised bed—'Little Sparty'—named after UNCG's Spartan mascot. Another faculty member from the Spanish Department enabled students to participate in growing: students had to label their beds in Spanish and speak to one another (and with me) in Spanish while we worked in the Gardens together. One long-time staff member/gardener grew interesting crops and adorned her beds with figurines. She said, 'Growing flowers and veg are like art, and therefore should be beautiful'. Her plots are frequently remarked upon, for even during the dormant winter months she displays fun objects; she is always changing them to keep them interesting. Without a doubt, the UNCG Gardens attract volunteers with a range of garden literacy and gardening experience. There are some students who have grown up in families that garden together, and there are those who have no gardening experience. Students share stories of their initial challenges when first participating at the garden. One student shared that while she was weeding with others, she 'didn't know what to wear', so she phoned her grandmother who gardens. Her grandma told her, 'to wear anything she didn't mind getting dirty'. In another case, a student was asked to transplant some plants from a multi-pack, which the student promptly did by digging a hole and placing the entire multi-pack, plastic and all, into the ground. Students talk among themselves about being afraid of spiders, snakes, and other bugs, while others look them up on insect apps and learn whether qualities are beneficial or harmful. After a couple of hours of working together, chatting about this and that, their fears are abated, and the work gets done.

To maintain community and a vibrant working group, the Student Garden Club meets twice a week for two-hour work sessions during the academic year. First-time attendees receive a pair of gardening gloves that they can keep; the gloves are meant to serve as an incentive to return. While the beginning of term has higher student participation than around exam or term-paper deadlines, there is a core group of regulars that rarely misses a week. During the winter months (January and February) students meet on Fridays for a couple of hours in a classroom and share tea or hot chocolate while working on garden projects, seed selection, and making plans for the garden. Students have used this time to make scarecrows and decorate bamboo labels with the names of the crops grown and other labels for tools. They have painted rocks to keep fencing and netting in place. As students pore over the seed catalogues together, we learn that some have never tasted a radish or a turnip or eaten a fresh carrot.

This element of teaching one another transcends the classroom and moves into the Gardens: it has been key to creating community. As Kalyn Milot, an anthropology major and past Social Media Coordinator, said:

> The gardens teaches skills, but I also think it teaches students that they are capable and are able to learn new things, which provides a sense of confidence and encourages them to try new things and learn in other areas. It sparks curiosity about the world around us.
>
> *(K. Milot (personal communication, 2025))*

These are points which speak to a wider audience, including practivists of various backgrounds.

There is a pace and a rhythm in the Gardens. In the process, volunteers who attend regularly learn what to plant seasonally. When they join in August and September, there is still a supply of summer vegetables to harvest. As the semester rolls on, the Gardens get winterised and prepared for spring planting. Each season is different: they vary according to when planting or harvesting can take place for certain crops, all of which undoubtedly has become more challenging with climate variability.

In the end, what has mattered most to sustaining the momentum at UNCG Gardens has been creating an inclusive and welcoming community—a space to be oneself. It is a place where differently abled bodies and minds and all genders, faiths, and languages come to work together. Embracing these differences allows us to learn from one another. A project conducted by Seth Rumbley, a former anthropology undergraduate and garden volunteer, questioned 30 students why they participated in the campus gardens. Interestingly, he learned that food security did not make the list from the responses generated (Rumbley 2016). Instead, students said they participate in the Student Garden Club to be together and to be part of a like-minded group while learning skills they want to have for their future. Students were interested in learning how to grow their own food at home or in a community garden, to care for the environment, but not to end hunger—for themselves or for others.[3] We also learned other students gardened to be closer to nature, spend productive time outside, and have meditative weeding sessions: it was their time for mindfulness. Students gardening over the summer did so to have access to the harvest because they valued eating and cooking with fresh produce.

Harvests and food sharing

In addition to serving as an educational tool for hands-on learning, the UNCG Gardens also serve as a local food source. When we gather the harvest, students may sample fresh produce and different herbs and take home what they would like. Cherry tomatoes do not often make it home to garnish a salad; they tend to be snacked on while the students garden. Some students wrap cherry tomatoes in basil leaves; we call them 'Sparty Poppers'. The regulars have learned to bring containers with them to be able to take home a share of the harvest. When we have a large harvest, the produce and herbs are donated to an on-campus food bank, the Spartan Open Pantry (SOP). This relationship has inspired volunteers at the SOP to volunteer at the student garden on workdays.

Over the years there have been bountiful harvests from the organic garden. Although the raised beds are small, we still manage to produce hundreds of pounds of fresh produce annually. As previously mentioned, the UNCG Gardens produces both for home consumption and for donations to local food banks and food pantries. Student volunteers who live on campus and often do not have their own cooking facilities and enjoy eating fresh from the garden that they can eat raw and occasionally take some fresh produce home to their parents. They also appreciate having an abundance of fresh produce to donate. In the past four years. The UNCG Gardens has donated well over 1200 lbs (544 kg) to the on-campus food bank. We recognise our efforts are not making a dent in solving the problem of on-campus food insecurity.[4] What we are doing is providing modest access to fresh produce on a weekly and seasonal basis to what the SOP is already providing for students and staff in need.

For a couple of pre-COVID seasons, we held a harvest potluck dinner in the fall for the Student Garden Club. Volunteers who were able to brought food to share. On two occasions, international exchange students who had been gardening with us brought food representing their home countries (France and Russia). During the winter months we also sponsored presentations by international students as well as some by UNCG students returning from study-abroad experiences to talk about food and culture. Frequently, these presentations were accompanied by food to sample, providing a 'taste' of food cultures in other countries,[5] which was an opportunity to reinforce the cultural nature of food.

Lessons learned

We have learned a few key lessons from our campus gardens, some of which we have already mentioned in this chapter. Other lessons learned include the wide range of support needed, including labour, funding, and varying degrees of experience among volunteers. One lesson learned is the need for support from a large array of people (see Figure 11.1).

University support comes by way of the land provided for the Garden as well as the water, legal services, and allowing staff and faculty to make time during the workday to toil in the garden. University carpenters built the much-needed storage shed, and the sustainability coordinator designed and set up the solar power charging station for the battery-operated power tools. PGL, however, paid for both the shed and solar panel setup, but needed approval for them from the university. Yet, there is more the university could do to support the Gardens, for example, by including it in campus strategic planning and providing regular funding for general operations, faculty release time, and maintenance.

Student support is critical both in terms of their voluntary labour, but also for their creative ideas. With the students' assistance, raised beds are maintained in and around them. Students also rent plots to grow food for themselves or to donate as part of service projects for courses. The Student Garden Club also plays a critical role in keeping the garden running throughout the year and by recruiting new members at campus events

Faculty support is helpful when they incorporate the gardens into their classes or through service-learning experience and accept the gardens as a location to complete a class assignment. Support is also welcomed when faculty rent bed(s), especially during the summer months.

Staff support is very important and comes from a wide variety of departments and offices: recycling, housing, carpentry, maintenance, senate, grounds, and others. Staff are also helpful when they rent beds over the summer months and participate in organised workdays.

And lastly, having *community support* in the form of monetary donations and supplies is critical. The UNCG Gardens have received wood, coffee grounds, soil, compost, mulch, and transplants and seed on an as-needed basis. Much information has been imparted to our gardeners by farmers and master gardeners over the years. Collectively, there are a lot of layers of support needed to maintain the garden going forward (see Figure 11.1).

Labour is always needed. A major labour challenge is the commitment to the raised beds over the summer months when students leave and temperatures can reach over 100° F (37° C). Many volunteers do not want to work during the hot, humid summer days. In the summer, the garden coordinator (me) ends up taking over the abandoned plots and donates the harvest to SOP. To manage the additional work, two students are hired for ten hours each a week for regular watering, weeding, harvesting, mowing, and whatever else needs doing.

Funding for projects is always needed. The UNCG Gardens has operated on a shoestring budget with careful stewardship of donations. Such funding goes to supplies and summer hires. Additionally, tools wear out, wood rots,

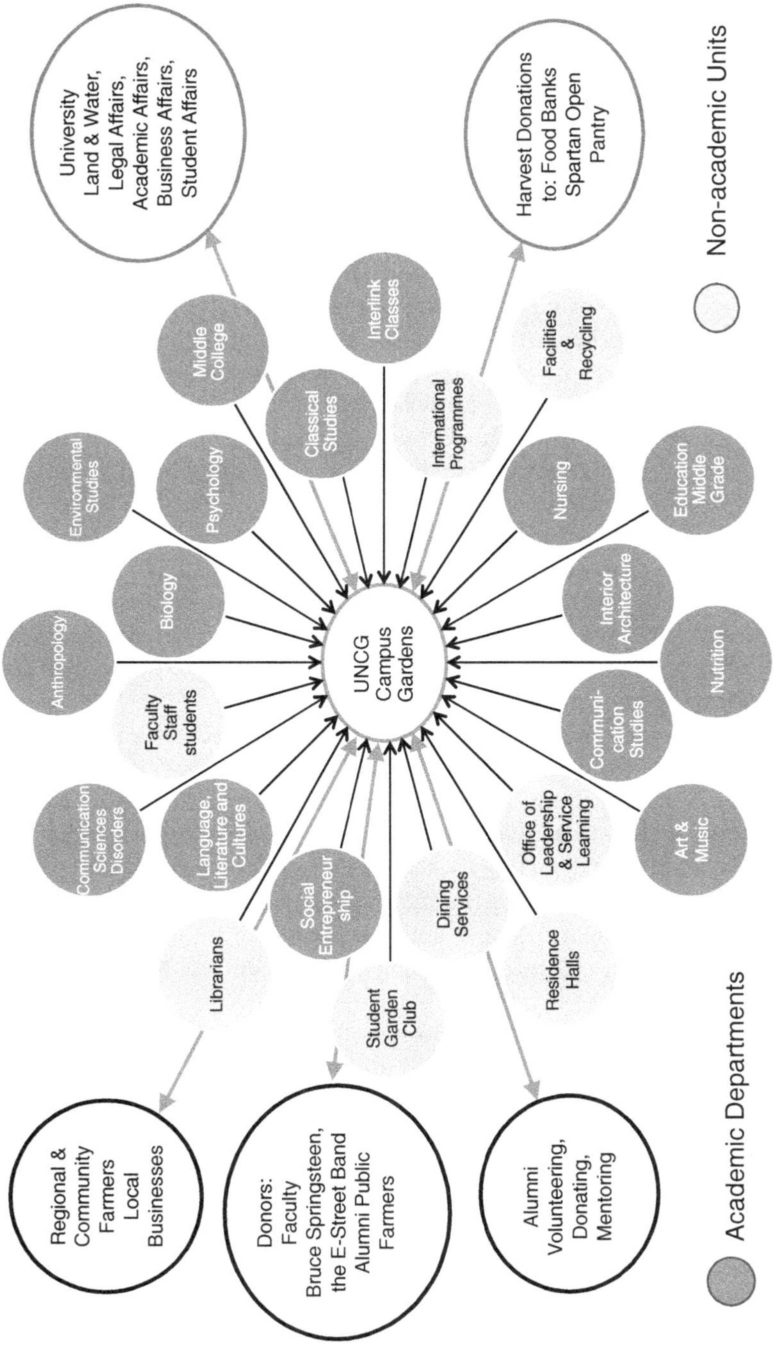

FIGURE 11.1 Sustaining an institutional green community at a campus garden

and new raised beds, benches, storage containers, and sheds need to be built, which takes an infusion of funds.

Speaking up about these needs is helpful. One afternoon I was lamenting the fact that the raised beds were breaking down after five years in the ground; they were returning to nature. A student asked what we needed, to which I replied: 'enough wood to replace all the beds'. Lo and behold, her father-in-law, the president of a lumber company, donated all the wood we needed. We used 'seconds' (not first quality), of course, but what a gift this was! On the third rebuild, a former student who had become a licensed contractor/builder used his professional discount to enable us to purchase the lumber at a reduced rate. He also brought his tools to the garden for a workday, trained women and men on power tool safety, and guided them in bench and raised bed construction. Most of the benches and a good many raised bed frames were assembled by women using power tools.[6]

Experience in gardening is important. As mentioned, we have learned there are mixed levels of gardening literacy among the garden volunteers. Volunteers, especially those new to gardening, would frequently ask what to plant and when. How far apart should the seeds be and how deep? How many plants or rows of seed should they put in a raised bed? When and how much to water? They also ask, 'Is this a weed and what kind is it?' One of the most frequently asked questions is: 'Will this bug hurt me?' Some will say they don't like garden critters (snakes, spiders, or pill bugs); others will pick up and relocate them to the compost pile.

Those new to gardening often do not know the names of tools or how to use them. Some students have never pushed a wheelbarrow, lifted a shovel, or worked a hoe. To help overcome the dilemma of 'which tool do I need', an indoor winter student activity was to make large name cards for the tools that also had a drawing depicting the tool. This signage is helpful for retrieval as well as for returning tools to the right place and maintaining a neat and tidy shed, especially when there are multiple users.

The Gardens is an exploratory learning environment. As Milot recounts:

> I learned a lot about gardening and myself by having to figure things out and learn what I wanted to know. How to ask questions, how to even know what questions to ask, how to try things and sometimes fail, and keep going and learn from my mistakes. It taught me a lot about how to teach, learn about my own impatience, and how to deal with that impatience. These are soft skills and are highly transferable.
>
> *(K. Milot (personal communication, 2025))*

These experiences illustrate the opportunity for critical learning outside the classroom. An important motivation for this process behind the scenes is to let students learn and fail in a low-stakes environment.

Conclusion: The impact of the UNCG Gardens

In my capacity as the UNCG Gardens co-director and Student Garden Club advisor these past 15 years, I have given this practivism plenty of food for thought. I hope this chapter has illustrated the possible social architecture of green communities from a grassroots perspective on a shoestring budget. To this end, the UNCG Gardens has become an educational learning environment fostering sustainability and community. Therefore, highlighting a few of its accomplishments seems appropriate here.

The UNCG Gardens has created community by virtue of its volunteer members; it has given them an identity tied to place, the campus Gardens, and an affiliation of practice, gardening together. Through the volunteer efforts noted above, our harvests have been donated to local food banks and churches. Moreover, since the Gardens was approved, over 4000 students have been part of the UNCG Gardens, nearly 50 faculty members from 15 departments, and countless numbers of staff from other non-academic units have contributed to its success (see Figure 11.1).

As a member of the Association for the Advancement of Sustainability in Higher Education (AASHE 2025), the university maintains a silver star sustainability rating, and the UNCG Gardens contributes to this rating (AASHE and STARS 2025). It is the participants, however, who gain important insights into gardening and sustainability, not the administration.

Garden skills have been shared with dedicated gardeners, and research has been conducted in various disciplines. The Student Garden Club is active and attracts ten to 35 students weekly. More recently, the dining hall and a residential dormitory have begun herb gardens, while other faculty members participate in community gardens elsewhere, at their churches and children's schools. Participation over the years has created an institutional green community and clearly demonstrates the viability of this type of community garden through repeated annual successes.

Although UNCG students reside in a food desert,[7] garden participants recognise that the Gardens in and of itself is not a sufficient source for fresh produce to feed themselves, the surrounding community, or supply SOP. However, it has become a setting for students to gain first-hand experience in gardening for their future needs—whether for personal satisfaction, better-tasting food, or creating friendships. It has become a place to learn about sustainability and to have first-hand experience of contributing to food justice. It is a community space and serves as a source of shared learning that provides a diverse, self-selected student body with experience, knowledge, and skills that will enable them to grow for themselves in the future and teach others should that opportunity arise. In other words, the Gardens is the first step of enabling them to become future practivists; as such, it reflects the scalable potential.

Once they graduate, some students seek out ways to stay connected to gardening and/or their local food system. A few students have gone on to participate in WWOOF International, and several have bought farms and now sell at local farmers' markets. A small number work part-time at farmers' markets or volunteer at local food pantries, while others work for Agricultural Extensions and Community Services. Some are teaching gardening in public schools, while still others participate in community gardening in their communities or grow their own home gardens with their children. Hence, the impact of the UNCG Gardens is lifelong and inter-generational.

As discussed above, we have learned various lessons from volunteering at the UNCG Gardens. Something special happens when people get outside, get their hands in the soil, and grow together. Clearly, the UNCG Gardens has become an ideal pedagogical tool for educating students about social justice, civic engagement, and personal responsibility for positive social change led by a practivist, a home gardener, and a 60-year local farmers' market supporter. More importantly, the UNCG Gardens has been instrumental in active learning about sustainability while growing fresh produce for local consumption for volunteers, those renting beds, or the campus food pantry. It is a safe space for relaxing and meditating while mowing or weeding. Those who get their hands dirty have benefitted from their experience and involvement. It is difficult to assess if the university and the current administration believe in the UNCG Gardens' future—we will have to see how they plan for its longevity.

Notes

1 The UNCG has an encompassing approach to sustainability (University of North Carolina Sustainability n.d.): 'Academics, operations, and outreach… conducted with careful attention to the enduring interconnectedness of social equality, the environment, the economy, and aesthetics' (UNCG Sustainability https://sustainability.uncg.edu/).

2 The UNCG Campus Gardens was named in the plural form to acknowledge each volunteer works in an individual raised bed. In the chapter the gardens are referred to as UNCG Gardens, the Gardens and UNCG Campus Gardens, as in uncggardens.uncg.edu

3 A few years back, one student volunteer who was graduating showed her father the Gardens on a family visit. She told him that when she graduates, leaving the Gardens and the friends she made there will be what she will miss the most about UNCG. Before she went on to graduate school, she chose to participate in World-Wide Opportunities on Organic Farms (WWOOF) as part of her gap time.

4 When you factor in the time from planting seeds and transplants to harvesting, a community garden is not a quick and easy solution to food insecurity and hunger. We might get a lot planted in the spring, but it takes at least 28 days before we see a radish and several months before we see a ripened tomato in late July. And if unwanted critters eat the plants, then the garden has less to offer. Hence,

gardening takes time; when one is hungry, waiting for a plant to mature is a long delay.

5 I would always say 'An eggplant is an eggplant or an aubergine, until you cook it. Then it can become Italian, Asian, Indian, or Middle Eastern cuisine'.

6 The UNCG Gardens was once located near a children's daycare centre, which has since moved. On one workday, children in their play area watched our volunteers construct the raised beds and fill them with soil. According to one of the child caretakers, 'One little boy didn't understand why all these girls were working with power tools—"that's what boys do like their daddy" '. At the garden, we don't genderise tools; everyone learns equally.

7 UNCG is located in a food desert. There is a plethora of fast-food restaurants and the dining hall within walking distance. However, the surrounding area of the campus is devoid of a major grocery store or supermarket within a three mile radius, which aligns with the USDA's definition of a food desert.

References

Andreatta, S. (2005). Urban connections to locally grown produce. In P. Barlett (Ed.), *Urban place: Reconnecting with the natural world* (pp. 117–140). The MIT Press.

Andreatta, S. (2006). When a good project goes awry: Community re-connecting with an urban farm. *Urban Anthropology*, 35(1), 75–104. www.jstor.org/stable/40553490

Andreatta, S. (2015). Through the generations: Victory gardens for tomorrow's tables. *Journal of Cultural, Agriculture, Food and Environment*, 37(1), 107–117. https://doi.org/10.1111/cuag.12046

Andreatta, S. (2022). Moving from literature to dialog to action: Teaching degrowth in anthropology courses. *Teaching & Learning Anthropology*, 4(2), 49–68. https://doi.org/10.5070/T34244393

Andreatta, S. (2025). Greening farmers markets: How communities come together to keep items and waste out of the landfill. In P. Kuppinger (Ed.), *Second-hand cultures and economies of reuse, repair, sharing, and care*. Palgrave Macmillan. https://doi.org/10.1007/978-3-031-99875-1

Andreatta, S., & Hoskins, M. (2022). Supporting a local agro-food system in COVID-19 times. *Anthropology and Ethnology*, 5(2), 000177. https://medwinpublishers.com/AEOAJ/supporting-a-local-agro-food-system-in-covid-19-times.pdf

Andreatta, S., Hoskins, M., Milot, K., & Vitale, L. (2023). Practicing anthropology: Fieldwork at its best—Community engagement through a university campus garden. *Practicing Anthropology*, 45(4), 32–38. https://doi.org/10.17730/0888-4552.45.4.32

Andreatta, S., Nash, B., & Martin, G. B. (2011). Carteret Catch™: Promoting local seafood through community and business partnerships. *Human Organization*, 70(3), 279–288. https://doi.org/10.17730/humo.70.3.l2284858q672j47g

Andreatta, S., & Parlier, A. (2010). The political ecology of small-scale commercial fishermen in Carteret County, North Carolina. *Human Organization*, 69(2), 180–191. https://doi.org/10.17730/humo.69.2.u41204x144664240

Andreatta, S., Rhyne, M., & Dery, N. (2008). Lessons learned from advocating CSAs for low-income and food insecure households. *Southern Rural Sociology*, 23(1), 1–33. https://egrove.olemiss.edu/jrss/vol23/iss1/6

Baker's Creek Heirloom Seed Company. (n.d.). Rare seeds. www.rareseeds.com

Blane, H., Sprowls, E. D., & Deskin, Z. (2004, 6 March). The campus garden initiatives can help grow the next generation of environmental change-makers. *The Conversation*. https://theconversation.com/campus-garden-initiatives-can-help-grow-the-next-generation-of-environmental-change-makers-222361

Bundy, H. (2024). Gardening as a mindfulness practice. *Extension Gardener*. https://extensiongardener.ces.ncsu.edu/2024/07/gardening-as-a-mindfulness-practice/

Centers for Disease Control and Prevention. (1997). *Principles of community engagement* (1st ed.). CDC/ATSDR Committee on Community Engagement.

Centers for Disease Control and Prevention. (n.d.). *Principles of community engagement* (2nd ed.). Centers for Disease Control. www.atsdr.cdc.gov/community-engagement/php/executive-summary/index.html

Community Places. (2014). Community planning toolkit. www.communityplanningtoolkit.org

Hoff, J., Gausset, Q., & Lex, S. W. (2020). Introduction. In J. Hoff, Q. Gausset, & S. W. Lex (Eds.), *The role of non-state actors in the green transition: Building a sustainable future* (pp. 1–13). Routledge.

Lawrence, A. (2022, 15 November). How university community gardens improve students' well-being. *Forest Nation Blog*. https://forestnation.com/blog/how-university-community-gardens-improve-students-well-being/?srsltid=AfmBOooJeVBjrbMj_K_0iTX9uURnIJjlLLoZ_cnfA03WQNlgJth6nkxV

March, P., Mallick, S., Flies, E., Jones, P., Person, S., Koolhof, I., Byrne, J., & Kendal, D. (2020). Trust, connection and equity: Can understanding context help to establish successful campus community gardens? *International Journal of Environmental Research and Public Health*, 17(20), 7476. https://doi.org/10.3390/ijerph17207476

McClosky, D. J., McDonald, M. A., Cook, J., Huertin-Roberts, S., Undergrove, S., Sampson, D., Gutter, S., & Eder, M. (2024, 12 November). Community engagement: Definitions and organizing concepts from the literature. *UC Davis Health Center for Regional Change*. https://health.ucdavis.edu/media-resources/crhd/documents/pdfs/principles-community-engagement-2nd-english.pdf

Meyer, A. (2022, 22 November). College community gardens promote growth and healing in COVID's wake. *EdSource*. https://edsource.org/2022/college-community-gardens-promote-growth-and-healing-in-covids-wake/681593

Murtagh, T. (2024, 9 September). The blossoming benefits of student gardens on college campuses. *Honors Society*. www.honorsociety.org/articles/blossoming-benefits-student-gardens-college-campuses

ostila, T. E., & Eriksson, C. (2024). Research praACTivism: A methodology for a more-than-educational academic knowledge production. *Gender and Education*, 37(1), 49–64. https://doi.org/10.1080/09540253.2024.2409894

Project Green Leaf. (2001). https://anthropology.uncg.edu/project-green-leaf/

Rumbley, S., & Andreatta, S. (2016). Forgotten farmers: Growing community in an urban food desert. In *Paper presented at the Society for Applied Anthropology Annual Conference*, 29 March–2 April, Vancouver, BC.

Spartan Open Pantry. (2023). Located at Wesley-Luther at UNC Greensboro. https://wesleyluther.org/sop

The Association for the Advancement of Sustainability in Higher Education (AASHE). (2025). https://aashe.org/

The Sustainability Tracking, Assessment & Rating System (ASHEE and STARS). (2025). A program of the association for the advancement of sustainability in higher education. https://stars.aashe.org/about-stars/

The University of North Carolina—Greensboro Campus Gardens. (n.d.). uncggardens.uncg.edu

University of North Carolina Sustainability. (n.d.). https://sustainability.uncg.edu/

United States Department of Agriculture. (2023). Census of agriculture impacts in the next generations. www.usda.gov/about-usda/news/blog/2023/02/22/2022-census-agriculture-impacts-next-generations-farmers

World-Wide Opportunities on Organic Farms (WWOOF). (n.d.). https://wwoof.net/

Yale University. Yale Sustainable Food Program. (2022). www.sustainablefood.yale.edu/the-yale-farm

CONCLUSION

Lessons learned from the social architecture of green communities

Anne-Kathrin Schwab, Quentin Gausset, Helen Jarvis, Simon Lex, Freja Bach Kristensen, Camilla Nielsen-Englyst

As this volume draws to a close, we reflect on the rich tapestry of green communities explored across its chapters, from Danish eco-villages and social housing estates to UK cohousing initiatives, Dutch urban gardens, and beyond. These case studies illuminate how alternative social organisations rooted in 'practivism' rather than mere activism foster sustainable living at the human scale.

'Practivism', as defined in the volume's introduction, emphasises changing one's own practices through collective action and represents a rupture from individualistic nudges or economic incentives. The chapters collectively demonstrate that green communities are not utopian enclaves but dynamic, imperfect experiments in social architecture, where interpersonal relations, decision making processes, and shared resources drive environmental and social impact.

The lessons learned in this book highlight similarities and differences among the cases. It addresses barriers to niche innovation in diverse cultural and national contexts, and offers recommendations for policy development and solutions in order to enhance scalability.

Drawing on the chapters' insights, we organise our reflections around three major questions:

(1) What do green communities bring to the green transition?
(2) How do they work?
(3) Can these models be dispersed?

In doing so, we bridge the volume's focus on Global North cases to broader implications that might also be relevant elsewhere.

DOI: 10.4324/9781003769064-13

What do green communities bring? Hope, human scale, trust, and practical solutions

Green communities offer (as shown) tangible contributions to sustainability, but their true value lies in rehumanising the green transition amid alienation, inequality, and environmental despair where technology seems on the level of economy and political decisions the answer to the poly-crisis, and green growth seems on the bigger scale still the only solution on answering environmental questions (cf. Jacobs 2013; Hobson 2013). The concept and paradigm that green communities follow seem to be on the other pole of the strategic decision making process for reaching the goals of sustainability. As the introduction to this book posits, these intentional groups—whether eco-villages, urban gardens, or renewable energy collectives—indeed reduce carbon footprints through shared economies of scale, permaculture, and behavioural change. Quantitative evidence from surveys cited in the introduction shows members' footprints are significantly below regional averages, which underscores their environmental impact.

Yet, this book's chapters reveal deeper social dividends. In chapter 1, Helen Jarvis examines UK people-powered green placemaking, where relational power—built through theories of change and interpersonal connections such as shared meals—fosters alliances for degrowth. This echoes the human-scale action in chapter 11, where Susan Andreatta describes a US university community garden that not only produces food but builds trust and equity through engagement, turning students into environmental stewards. Similarly, Freja Bach Kristensen's chapter 9 on social housing in Denmark highlights how sustainability brokering by individuals aligns green activities with residents' life projects and yields higher life satisfaction as well as increased resource awareness.

These examples illustrate hope as a practical outcome: green communities provide spaces for engaging in alternative future making by practising engagements between people and the planetary ecosystems that sustain us and build on care and mutual benefit. This confirms the potential of prefigurative politics as played out in community activities to enable solidarity and critical engagement with societal conventions (Cooper 2020; Nelson & Chatterton 2022; Chatterton et al. 2013). This is further highlighted in Cristina Grasseni's chapter 10 on Dutch urban gardens, which portrays 'slow practivism'—volunteering that emphasises sensory learning, seasonality, and conviviality—as a form of resistance to efficiency-driven narratives that rebuilds trust through embodied skills. Ieva Snikersproge's typology in chapter 4 of French back-to-the-land models—from communes to eco-habitats—shows how subsistence practices enhance autonomy, offering practical solutions like relocalised production that counter capitalist time disciplines.

At their core, green communities restore human-scale collaboration where people can meaningfully cooperate and live true relationships. They address inequality by decoupling well-being from wealth, as seen in Penny Clark's chapter 7 on UK intentional communities, where sharing spaces and objects reconceptualises 'home' as communal, reducing emissions from goods and heating. This relational work, as Ana Margarida Esteves argues in chapter 2 in her analysis of intentional communities Tamera and Monte Verità, deploys post-patriarchal soft power—values of care and harmony—to attract support, though it risks reproducing class barriers. It has to be clearly stated, as Esteves does, that green communities are not the answer to all problems. Trauma and pain can be reproduced and made even more obvious in close settings. However, with awareness and a careful focus on this, it becomes possible to change the social architecture of green communities. These communities can become islands of healing, but this requires a high reflexivity of all members and a current supervision and trauma-sensitive environment and inter-environment (cf. Burschel 2020).

While green communities are often studied through the angle of spirituality (e.g., Farkas 2021; Koziol 2020; Brito 2023; Pinto 2024), the contributions to this book focus more on the pragmatic work of finding ways of living more sustainably and of keeping communities together within adequate social architectures. The cases presented in this volume all display examples of strong 'spiritual capital', understood as the ability to imagine and create new worlds (Eisenstein 2007; 2011). But the creation of alternative ways of living and organising is analysed here as profane and mundane rather than as sacred.

In sum, green communities bring hope not through romanticism but through pragmatic anticapitalist critiques (whether explicitly or implicitly) and a high self-reflexivity: they model just transitions that integrate social justice, as in Simon Lex's chapter 5 on Danish 'constructive disobedience', where resistance to top-down energy projects evolves into community-led alternatives that enhance cohesion.

How do green communities work? Cohesion, agile decision making, and addressing challenges

Green communities thrive or falter based on their social architecture—that is, how they manage conflicts, make decisions, and navigate inclusion/exclusion. The chapters here reveal that success demands agile, inclusive processes that balance individual autonomy with collective goals while confronting darker issues like exclusion or burnout.

Decision making emerges as a key theme. In chapter 3, Quentin Gausset and Camilla Nielsen-Englyst contrast majority-rule associations with sociocracy and do-ocracy in Danish green communities, showing how consent-based models in eco-villages and beekeeper guilds foster efficiency and trust without coercion. This agility is mirrored in Anne-Kathrin Schwab's chapter 6 on transforming German rurban areas, where participatory methods from eco-villages—like Dragon Dreaming, Theory U and sociocratic circles—build cohesion in a village in Northern Lower Saxony, Germany by respecting traditional structures while sparking sustainability actions. These chapters demonstrate how relevant the experiences and social experiments of green communities are in order to be able to transform habits and attitudes of democratic practices on a local scale and foster sensitivity, humanity, responsibility and connectedness between individuals.

Yet, cohesion requires the addressing of exclusion. Esteves (chapter 2) warns that 'countercultural soft power' can create barriers for those lacking class privilege or cultural fluency, as in the Tamera community's unpaid trials. Snikersproge (chapter 4) discusses decollectivisation pressures in French models, where communes evolve into eco-habitats due to capitalist influences, which highlights the need for ethical grounding to avoid far-right co-optation. Cristina Grasseni (chapter 10) critiques top-down efficiency in Dutch gardens, where volunteers' experiential learning clashes with managerial expectations and leads to disengagement.

Interpersonal and relational work is crucial. Helen Jarvis (chapter 1) emphasises 'relational togetherness' through theories of change, while Penny Clark (chapter 7) details negotiations of privacy norms in shared spaces that require skills in conflict resolution such as nonviolent communication. Freja Bach Kristensen (chapter 9) underscores sustainability brokers' role in translating agendas, but notes their dependence on collaboration, acknowledgment, and resources to avoid burnout.

Challenges abound: Susan Andreatta (chapter 11) shows how university gardens face institutional hurdles but succeed through the presence of student voices and funding. Simon Lex (chapter 5) highlights mistrust caused by unequal energy project burdens, advocating the presence of 'experts by lived experience' in planning projects. Overall, communities work when they are open yet cohesive and agile yet ethical—and reinforce each other through interconnected elements like governance, brokering, and hope.

How can these models be spread? Scaling up through decentralisation, support, and multi-stakeholder engagement

Scalability is this volume's ultimate challenge: how to multiply green communities without diluting their essence? As outlined in the introduction, there can be contradictions between scaling up and out, i.e., replicating or proliferating core principles and approaches, and scaling deep and within, i.e., focusing on creating long-lasting and durable change. The chapters avoid 'scaling up' in a productivist sense. Echoing degrowth arguments, the answer to these questions lies rather in keeping the scale of green communities down and in scaling them out, decentralising, localising, and multiplying humanistic models while addressing barriers. The main dilemma here is how to remain attentive to local needs and conditions to foster senses of belonging and meaningful engagements while also framing community activities within a wider transformational agenda where impact lies beyond the immediate community activities (Aiken et al. 2022).

Decentralisation is key. Snikersproge (chapter 4) argues that relocalising food production multiplies smaller communities, as in French eco-habitats. However the focus on the values of the intentional communities that compete clearly with the values of right-wing local initiatives have to be reflected very carefully in each community. Grasseni (chapter 10) advocates slow resistance over rapid system-change, preserving skills like crafts, taste, and conviviality. Schwab (chapter 6) demonstrates how action research could actually scale eco-village methods to rurban areas, fostering participatory transformations, but only if it is practiced with sensitivity and if it respects the traditions of the locals. Thus, scaling happens through spreading methods, attitudes, and ideas.

Multi-stakeholder engagement unlocks barriers. Lex (chapter 5) calls for municipalities to act as intermediaries by engaging informal groups in energy planning. Bach Kristensen (chapter 9) illustrates that social housing's infrastructures support brokering but require alignment with residents' ambitions and governmental resources. Jarvis (chapter 1) highlights alliances with authorities for placemaking, while Andreatta (chapter 11) leverages university funding and student involvement in gardens.

Policy pointers for spreading green communities include: supporting brokers with training and funds, regulating common governance (Esteves, chapter 2), and fostering intersectional solidarity. Local governments can provide land, as in Dutch gardens, or mediate dialogues, as in German rurban projects.

Yet, scaling demands pragmatism amid a poly-crisis. Northern European focus notwithstanding, these cases contribute to global activism by modelling resilience against inequality and authoritarianism. Further research should connect Global North niches to Global South crises to explore hybrid models (cf. Roysen et al. 2026).

Wider significance: Provocation for activism in a poly-crisis

This volume's cases—predominantly from Europe and North America—address potential concerns about niche, middle-class biases by emphasising pragmatic anti-capitalism: green communities as intermediaries in just transitions, not as isolated utopias. Green communities provoke a question: in an era of grief and grievances, can we reclaim community as a bulwark against division? Ethical localisation risks exclusion, but inclusive brokering offers hope. However, knowledge and wisdom from indigenous communities and the Global South are open for further discussion. As Eisenstein (2011) says, without a deep inner change of attitude and spiritual capital, the transformation of sustainability cannot succeed.

Ultimately, green communities remind us that sustainability is relational. Collective action ruptures individualism. In a poly-crisis, green communities seed engagement and hope—practical, human-scale, and interconnected hope—and pave ways to equitable, resilient futures. This is our call: not just to study, but to support on any level (politically, academically, locally) the practices and activities of true 'practivists' (Hoff et al. 2020) because this is what we can do for this world as observers, thinkers, and citizens.

References

Aiken, G. T., Eadson, W., Hobson, K. & Dinnie, L. (2022). Vicarious scale and instrumental imaginaries in community sustainable transitions. *Global Environmental Change*, 75, 102543. https://doi.org/10.1016/j.gloenvcha.2022.102543

Brito, L. G. (2023). *Ecological epistemologies and spiritualities in Brazilian ecovillages: In the labyrinth of an environmental anthropology*. Routledge.

Burschel, M. (2020). *Trennung in Verbundenheit?* [Separation in Unity]. [Doctoral dissertation, Ludwig Maximilian University of Munich].

Chatterton, P., Featherstone, D., & Routledge, P. (2013). Articulating climate justice in Copenhagen: Antagonism, the commons, and solidarity. *Antipode*, 45(3), 602–620.

Cooper, D. (2020). Towards an adventurous institutional politics: The prefigurative 'as if' and the reposing of what's real. *The Sociological Review*, 68(5), 893–916.

Eisenstein, C. (2007). *The ascent of humanity*. Panenthea Press.

Eisenstein, C. (2011). *Sacred economics: Money, gift & society in the age of transition*. North Atlantic Books.

Farkas, J. (2021). Religion and spirituality in Hungarian eco-villages. In S.E. Silvern, & E.H. Davis (Eds), *Religion, sustainability, and place: Moral geographies of the Anthropocene* (pp. 225–251). Palgrave Macmillan,. https://doi.org/10.1007/978-981-15-7646-1_10

Hobson, K. (2013). 'Weak' or 'strong' sustainable consumption? Efficiency, degrowth, and the 10-year framework of programmes. *Environment and Planning C: Government and Policy*, 31(6), 1082–1098. https://doi-org.kb-ku.idm.oclc.org/10.1068/c12279

Hoff, J., Gausset, Q., & Lex, S. W. (2020). Introduction. In J. Hoff, Q. Gausset, & S. W. Lex (Eds.), *The role of non-state actors in the green transition. Building a sustainable future* (pp. 1–13). Routledge.

Jacobs, M. (2013). Green growth. In R. Falkner (Ed.), *The handbook of global climate and environment policy* (pp. 197–214). Wiley-Blackwell.

Koziol, C. A. (2020). *An exploration of four ecovillages through the ecoresilient lens of spirituality*. [Doctoral dissertation, Laurentian University].

Nelson, A., & Chatterton, P. (2022). Dwelling beyond growth: Negotiating the state, mutualism and commons. In F. Savini, A. Ferreira, & K. von Schönfeld (Eds.), *Post-growth planning: Cities beyond the market economy* (pp. 49–62). Routledge.
Pinto, T. (2024). Spiritual representations in rural ecovillages: A systematic review. *Journal for the Study of Religion, Nature & Culture*, 18(2), 278–303.
Roysen, R., Ayala, D. H. F., Ramos-Mejía, M., Kos, L., Bruehwiler, N., & Koehrsen, J. (2026). Replicating grassroots innovations in the Global South: A case study of ecovillages for marginalized groups in Colombia. *Environmental Innovation and Societal Transitions*, 58, 1–16. Article 101068.

INDEX

Note: Page numbers in *italics* indicate figures and in **bold** indicate tables on the corresponding pages.

For Product Safety Concerns and Information please contact our EU representative GPSR@taylorandfrancis.com
Taylor & Francis Verlag GmbH, Kaufingerstraße 24, 80331 München, Germany

www.ingramcontent.com/pod-product-compliance
Lightning Source LLC
LaVergne TN
LVHW020802241026
850861LV00034B/449
* 9 7 8 1 0 4 1 2 9 3 2 9 3 *